ANNALS OF MATHEMATICS STUDIES
NUMBER 7

FINITE DIMENSIONAL VECTOR SPACES

BY

PAUL R. HALMOS

PRINCETON
PRINCETON UNIVERSITY PRESS
LONDON: HUMPHREY MILFORD
OXFORD UNIVERSITY PRESS
1948

Lithoprinted in U.S.A.
EDWARDS BROTHERS, INC.
ANN ARBOR, MICHIGAN

PREFACE

That Hilbert space theory and elementary matrix theory are intimately associated came as a surprise to me and to many colleagues of my generation only after studying the two subjects separately. This is deplorable: it took us as much time to discover for ourselves that there is a connection as it took to learn the two seemingly separate disciplines. I present this little book in an attempt to remedy the situation. Addressing the advanced undergraduate or beginning graduate student, I treat linear transformations on finite dimensional vector spaces by the methods of more general theories. My purpose is to emphasize the simple geometric notions common to many parts of mathematics and its applications, and to do this in a language which gives away the trade secrets and tells the student what is in the back of the minds of people proving theorems about integral equations and Banach spaces. The reader does not, however, have to share my prejudiced motivation. Except for an occasional reference to undergraduate mathematics the book is self contained and may be read by anyone who is trying to get a feeling for the linear problems usually discussed in courses on "matrix theory" or "higher algebra". The algebraic, coordinate - free, methods do not lose power and elegance by specialization to a finite number of dimensions, and are, in my belief, as elementary as the classical coordinatized treatment.

I originally intended this book to contain a theorem if and only if an infinite dimensional generalization of it already exists. Barring a few concessions to the tempting easiness of some essentially finite dimensional notions and results, I have followed this plan. My emphasis, however, is more on method than on results. The reader may sometimes see some obvious way of shortening the proofs I give. (He is, for example, very likely to do this in connection with the representation of a linear functional by an inner product or the treatment of direct products of unitary spaces.) The chances are that the infinite dimensional analog of the shorter proof is either much longer or else non existent.

To supplement the hints in the body of the book concerning the various directions in which a student may proceed, I have appended a bibliography. This very short list makes no pretense to completeness; it consists merely of the books which have helped me the most. Their perusal should give the student an idea of most of the important extensions of the subjects I treat.

In conclusion I want to express my really sincere thanks to virtually every mathematician in Princeton. Most of them have read parts of the manuscript, discussed the project with me, and were very kind in giving encouragement and criticism. I am particularly grateful to two men: John von Neumann, who is one of the originators of the modern spirit and methods which I have tried to present and whose teaching was the inspiration for this book, and J. L. Doob, who read the entire manuscript and made many valuable suggestions.

Paul R. Halmos

The Institute for Advanced Study,
Princeton, New Jersey

TABLE OF CONTENTS

ERRATA

p. 24, line 11: instead of $x_0(y)$ read $z_0(y)$.

p. 28, line 15: instead of $\mathfrak{U} \oplus \mathfrak{W}$ read $\mathfrak{U} \oplus \mathfrak{B}$.

p. 28, line 18: instead of $\mathfrak{W}$ read $\mathfrak{B}$.

p. 55, line 18: instead of $\mathfrak{M}_1 \cap \mathfrak{M}_2$ read $\mathfrak{M}_1 \cap \mathfrak{N}_2$.

p. 74, line 1: instead of $(AB \cdot B^{-1})$ read $\rho(AB \cdot B^{-1})$.

p. 97, line 9: instead of (7) read (1).

p. 122, last line: instead of (A + il) read (A + il)y.

p. 137: The first two sentences of the paragraph beginning near the bottom of the page should read as follows: Using the above characterization of non negativeness, the reader may verify that if $A = \begin{bmatrix} 1 & 0 \\ 0 & 0 \end{bmatrix}$ and $B = \begin{bmatrix} 0 & 0 \\ 0 & 1 \end{bmatrix}$, and if C is a Hermitian matrix for which both $A \leqq C$ and $B \leqq C$, then

$$C = \begin{bmatrix} 1 + \epsilon & \theta \\ \bar{\theta} & 1 + \delta \end{bmatrix},$$

where $\epsilon \geqq 0$, $\delta \geqq 0$, and $|\theta|^2 \leqq \min \{\epsilon(1 + \delta), \delta(1 + \epsilon)\}$. It is also easy to show that, for a matrix of the type of C, $C \leqq 1$ can hold if and only if $C = 1$.

1948 reprinting

Chapter I

SPACES

§1.

In what follows we shall have occasion to use different classes of numbers (such as the class of all real numbers or the class of all complex numbers). Because we don't want, at this early stage, to commit ourselves to any specific class we shall adopt the dodge of referring to numbers as *scalars*. The reader will not lose anything essential if he consistently interprets scalars as real numbers or as complex numbers: in the examples that we shall give both classes will occur.

DEFINITION. A *vector space*, $\mathfrak{V}$, is a set of elements x, y, z, etc., called *vectors*, satisfying the following axioms.

A.

To every pair x and y, of vectors in $\mathfrak{V}$ there corresponds a vector z, called the *sum* of x and y, z = x+y, in such a way that

(1) addition is commutative, x+y = y+x;

(2) addition is associative, x+(y+z) = (x+y) + z;

(3) there exists in $\mathfrak{V}$ a unique vector, 0, (called the *origin*) such that for all x in $\mathfrak{V}$, x+0 = x; and

(4) to every x in $\mathfrak{V}$ there corresponds a unique vector, denoted by -x, with the proper-

ty $x + (-x) = 0$.

B.

To every pair, α and x, where α is a scalar and x is a vector in $\mathfrak{V}$, there corresponds a vector y in $\mathfrak{V}$, called the <u>product</u> of α and x, $y = \alpha x$, such that

(1) multiplication is distributive with respect to vector addition, $\alpha(x+y) = \alpha x + \alpha y$;

(2) multiplication is distributive with respect to scalar addition, $(\alpha + \beta)x = \alpha x + \beta x$

(3) multiplication is associative, $\alpha(\beta x) = (\alpha\beta)x$; and

(4) $0x = 0$, $1x = x$.

(These axioms are not logically independent: they are merely a convenient characterization of the objects we wish to study.) According as scalars are interpreted as real or complex numbers we shall refer to real or complex vector spaces.

§2. EXAMPLES OF VECTOR SPACES

Before discussing the implications of these axioms we give some examples. We shall refer to these examples over and over again and we shall use the notation established here throughout the rest of our work.

(1) Let $\mathfrak{C}_1$ be the set of all complex numbers; if we interpret $x+y$ and αx as ordinary complex numerical addition and multiplication, $\mathfrak{C}_1$ becomes a complex vector space.

(2) Let $\mathfrak{P}$ be the set of all polynomials with <u>complex</u> coefficients in a <u>real</u> variable t. (There is no deep reason for this arbitrary choice: it is merely a matter of convenience for the purpose of giving examples later). To make $\mathfrak{P}$ into a complex vector space we interpret addition and scalar multiplication as the ordina-

ry addition of two polynomials and multiplication of a polynomial by a complex number, respectively; the origin in $\mathfrak{P}$ is the polynomial identically zero.

Example (1) is too simple and example (2) too complicated to be typical of the main contents of this book. We give now another example of complex vector spaces which (as we shall see later) is general enough for all our purposes.

(3) Let $\mathfrak{C}_n$, n = 1, 2, ..., be the set of all n-tuples of complex numbers, $x = \{\xi_1, \ldots, \xi_n\}$; if $y = \{\eta_1, \ldots, \eta_n\}$ we write, *by definition*,

$$x+y = \{\xi_1 + \eta_1, \ldots, \xi_n + \eta_n\},$$
$$\alpha x = \{\alpha\xi_1, \ldots, \alpha\xi_n\},$$
$$0 = \{0, \ldots, 0\}$$
$$-x = \{-\xi_1, \ldots, -\xi_n\}.$$

It is easy to verify that all parts of our two axioms (A) and (B), §1, are satisfied, so that $\mathfrak{C}_n$ is a complex vector space; it is usually called *n-dimensional complex Euclidean space*.

(4) For any positive integer n let $\mathfrak{P}_n$ be the set of all polynomials (with the same restrictions as in (2)) of degree $\leqq$ n-1, together with the polynomial identically zero. (In the usual discussion of degree the degree ~~the degree~~ of this polynomial is not defined, so that we cannot say that it has degree $\leqq$ n-1.) With the same interpretation of the linear operations (addition and scalar multiplication) as in (2), $\mathfrak{P}_n$ is a complex vector space.

(5) A close relative of $\mathfrak{C}_n$ is the set $\mathfrak{R}_n$ of all n-tuples of *real* numbers, $x = \{\xi_1, \ldots, \xi_n\}$. With the same formal definitions of addition and scalar multiplication as for $\mathfrak{C}_n$, excepting that we consider only *real* scalars α, the space $\mathfrak{R}_n$, the *ordinary* or real *n-dimensional Euclidean space*, is a real vector space.

§3. COMMENTS ON NOTATION AND TERMINOLOGY

A few comments on our axioms and notation. Those familiar with algebraic terminology will have recognized the axioms (A), §1, as the defining conditions of an abelian (commutative) group; the axioms (B) express the fact that the group admits scalars as operators. We use the "scalar" terminology to emphasize the fact that we are not even necessarily dealing with real or complex numbers. Ninety percent of the theory remains valid if we interpret scalars as elements of any field. If scalars are elements of a ring a vector space is sometimes called a modul.

Special real vector spaces (namely $\mathfrak{R}_n$) are familiar in geometry. There seems at this stage to be no excuse for the apparently uninteresting insistence on complex numbers. We hope that reader is willing to take it on faith that we shall have to make use of deep properties of complex numbers later, (conjugation, algebraic closure), and that in both the applications of vector spaces to modern (quantum mechanical) physics and in the mathematical generalization of our results to Hilbert space, complex numbers play an important role. Their one great disadvantage is the difficulty of drawing pictures: the ordinary picture (Argand diagram) of $\mathfrak{C}_1$ is indistinguishable from that of $\mathfrak{R}_2$, and a graphic representation of $\mathfrak{C}_2$ seems to be out of human reach. On occasions when we have to use pictorial language we shall therefore use the terminology of $\mathfrak{R}_n$ in $\mathfrak{C}_n$, and speak of $\mathfrak{C}_2$, for example, as a plane.

Finally we comment on notation. We observe that the symbol 0 has been used in two meanings: once as a number and once as a vector. To make the situation worse we shall later, when we introduce linear functionals and linear transformations, give it still other meanings. Fortunately the relation between the various interpretations of 0 is such that after this word of warning no confusion should arise from this practice. Another notationally happy circumstance is that $-x$ (defined in §1, (A)(4)) and $(-1)x$ are the same thing. This is true since

$$x + (-1)x = 1x + (-1)x = (1 + (-1))x = 0x = 0.$$

§4. DEFINITION OF LINEAR DEPENDENCE

Now that we have described the spaces we shall work with we must specify the relations among the elements of these spaces which will be of interest to us. Vector spaces are used to study linear problems: the general form of a linear relation is described in the following definition.

DEFINITION. A finite set of vectors, $x_1, \ldots, x_n$, is <u>linearly dependent</u> if there exist scalars $\alpha_1, \ldots, \alpha_n$, not all zero, such that

$$\sum_i \alpha_i x_i = \alpha_1 x_1 + \cdots + \alpha_n x_n = 0.$$

If, on the other hand, $\sum_i \alpha_i x_i = 0$ implies that $\alpha_1 = \cdots = \alpha_n = 0$, the vectors $x_1, \ldots, x_n$ are <u>linearly independent</u>.

Linear dependence or independence are properties of <u>sets</u> of vectors; it is customary however to apply these adjectives to vectors themselves and thus we shall sometimes say "a set of linearly independent vectors" instead of "a linearly independent set of vectors". It will be convenient also to speak of the linear dependence or independence of a not necessarily finite set, $\mathfrak{X}$, of vectors. We shall say that $\mathfrak{X}$ is linearly independent if every finite subset of $\mathfrak{X}$ is such; otherwise $\mathfrak{X}$ is linearly dependent.

To gain insight into the meaning of linear dependence let us study the examples of vector spaces that we already have.

(1) If x and y are any two vectors in $\mathfrak{C}_1$, then x and y form a linearly dependent set. If $x = y = 0$ this is trivial; if not we have, for example, the relation $yx + (-x)y = 0$. Since it is clear that any set containing a linearly dependent subset is itself linearly dependent, this shows that any set containing more than one element is a linearly dependent set.

(2) More interesting is the situation in the space $\mathfrak{P}$. The vectors $x = x(t) = 1 - t$, $y = y(t) = t(1 - t)$, and $z = z(t) = 1-t^2$ are, for example, linearly dependent, since $x + y - z = 0$. However the infinite set of vectors $x_0, x_1, x_2, \ldots$ defined by

$x_0(t) = 1$, $x_1(t) = t$, $x_2(t) = t^2$, $x_3(t) = t^3$, ...,

is a linearly independent set, for if we had any relation of the form

$$\alpha_0 x_0 + \alpha_1 x_1 + \cdots + \alpha_n x_n = 0,$$

then we should have a polynomial relation

$$\alpha_0 + \alpha_1 t + \alpha_2 t^2 + \cdots + \alpha_n t^n \equiv 0,$$

whence $\alpha_0 = \alpha_1 = \cdots = \alpha_n = 0$

(3) As we mentioned before, the spaces $\mathfrak{C}_n$ are the prototype of what we want to study: let us examine, for example, the case $n = 3$. To those familiar with higher dimensional geometry the notion of linear dependence in this space (or, more properly speaking, in its real analog $\mathfrak{R}_3$) has a concrete geometric meaning which we shall only mention. In geometrical language: two vectors are linearly dependent if and only if they are collinear with the origin, and three vectors are linearly dependent if and only if they are coplanar with the origin. (If one thinks of a vector not as a point in a space but as an arrow pointing from the origin to some given point, the preceding sentence should be modified by crossing out the phrase "with the origin" both times that it occurs). We shall presently introduce the notion

of linear manifolds (or vector subspaces) in a vector space and use the geometrical language thereby suggested.

For a concrete example consider the vectors $x = \{1, 0, 0\}$, $y = \{0, 1, 0\}$ $z = \{0, 0, 1\}$, and $u = \{1, 1, 1\}$. These four vectors form a linearly dependent set, since $x + y + z - u = 0$; it is easy to verify however that any three of these vectors form a linearly independent set.

§5. CHARACTERIZATION OF LINEAR DEPENDENCE

Returning to the general considerations we shall say, whenever $x = \alpha_1 x_1 + \cdots + \alpha_n x_n$, that x is a <u>linear combination</u> of $x_1, \ldots, x_n$; we shall use without any further explanation all the simple grammatical implications of our terminology. Thus we shall say, in case x is a linear combination of $x_1, \ldots, x_n$, that x is linearly dependent on $x_1, \ldots, x_n$; we shall leave to the reader the proof of the fact that if $x_1, \ldots, x_n$ are linearly independent then x is a linear combination of them if and only if the vectors $x, x_1, \ldots, x_n$ are linearly dependent.

The fundamental theorem concerning linear dependence is the following.

THEOREM. The set of non zero vectors, $x_1, \ldots, x_n$, is linearly dependent if and only if some x_k, $2 \leq k \leq n$, is a linear combination of the preceding ones.

PROOF. Let us suppose that the set $x_1, \ldots, x_n$ is linearly dependent and let k be the first integer between 2 and n for which $x_1, \ldots, x_k$ is linearly dependent. (If worse comes to worst the hypothesis of the theorem assures us that $k = n$ will do). Then

$$\alpha_1 x_1 + \cdots + \alpha_k x_k = 0$$

for a suitable set of α's; moreover, whatever the α's,

$\alpha_k = 0$ is impossible, for then we should have a linear dependence relation among $x_1, \ldots, x_{k-1}$, contrary to the definition of k. Hence

$$x_k = \left[\frac{-\alpha_1}{\alpha_k}\right] x_1 + \cdots + \left[\frac{-\alpha_{k-1}}{\alpha_k}\right] x_{k-1},$$

as was to be proved. This proves the necessity of our condition; sufficiency is clear since, as we remarked before, any set containing a linearly dependent set is itself such.

§6. DEFINITION AND CONSTRUCTION OF BASES

DEFINITION. A (linear) <u>basis</u> (or a <u>coordinate system</u>) in a vector space $\mathfrak{V}$ is a set $\mathfrak{X}$ of linearly independent vectors such that every vector in $\mathfrak{V}$ is a linear combination of elements of $\mathfrak{X}$. A vector space $\mathfrak{V}$ is <u>finite dimensional</u> if it has a finite basis.

Except for the occasional consideration of examples we shall restrict our attention, throughout this book, to finite dimensional vector spaces.

For examples of bases and finite and non finite dimensional spaces we turn again to the spaces $\mathfrak{C}_n$ and $\mathfrak{P}$. In $\mathfrak{P}$ the set $x_n = x_n(t) = t^n$, $n = 0, 1, 2, \ldots$ is a basis: every polynomial is, by definition, a linear combination of a finite number of x_n. Moreover $\mathfrak{P}$ has no finite basis, for given any finite set of polynomials we may find a polynomial of higher degree than any of them: this latter polynomial is obviously not a linear combination of the former ones.

An example of a basis in $\mathfrak{C}_n$ is the set of vectors x_i, $i = 1, \ldots, n$, defined by the condition that the j-th coordinate of x_i is δ_{ij}. (Here we use for the first time the popular Kronecker δ; it is defined by $\delta_{ij} = 1$ if $i = j$, and $\delta_{ij} = 0$ if $i \neq j$). Thus we assert that in $\mathfrak{C}_3$ the vectors $x_1 = \{1, 0, 0\}$, $x_2 = \{0, 1, 0\}$, and $x_3 = \{0, 0, 1\}$ are a basis. We have seen before that they are linearly independent; the formula

$$x = \{\xi_1, \xi_2, \xi_3\} = \xi_1 x_1 + \xi_2 x_2 + \xi_3 x_3$$

proves that every x in $\mathfrak{C}_3$ is a linear combination of them.

In a general finite dimensional vector space $\mathfrak{V}$, with basis $x_1, \ldots, x_n$, we know that every x may be written in the form

$$x = \sum_i \xi_i x_i;$$

we assert that the ξ's are uniquely determined by x. The proof of this assertion is an argument often used in the theory of linear dependence. If we had $x = \sum_i \eta_i x_i$, then we should have, by subtraction,

$$\sum_i (\xi_i - \eta_i) x_i = 0.$$

Since the x's are linearly independent, this implies that $\xi_i - \eta_i = 0$ for $i = 1, \ldots, n$: in other words the η's are the same as the ξ's.

THEOREM. If $\mathfrak{V}$ is a finite dimensional vector space and $y_1, \ldots, y_m$ is any set of linearly independent vectors in $\mathfrak{V}$, then, unless the y's already form a basis, we can find vectors $y_{m+1}, \ldots, y_{m+p}$ so that the totality of y's, $y_1, y_2, \ldots, y_m, y_{m+1}, \ldots, y_{m+p}$, forms a basis. In other words: every linearly independent set can be extended to a basis.

PROOF. Since $\mathfrak{V}$ is finite dimensional it has a finite basis, say $x_1, \ldots, x_n$. We consider the set $\mathfrak{S}$ of vectors

$$y_1, \ldots, y_m, x_1, \ldots, x_n,$$

in this order, and we apply to this set the theorem of §5 several times in succession. In the first place the set $\mathfrak{S}$ is linearly dependent since the y's are (as are all vectors) linear combinations of the x's. Hence some vector of $\mathfrak{S}$ is a linear combination of the preceding ones; let z be the first such vector. Then z is different from any y_i, $i = 1, \ldots, m$, (since the y's are linearly independent), hence z is equal to some x, say $z = x_i$. We consider the new set $\mathfrak{S}'$ of vectors

$$y_1, \ldots, y_m, x_1, \ldots, x_{i-1}, x_{i+1}, \ldots, x_n.$$

We observe that every vector in $\mathfrak{V}$ is a linear combination of vectors in $\mathfrak{S}'$ since by means of $y_1, \ldots, y_m$, $x_1, \ldots, x_{i-1}$ we may express x_i, and then by means of $x_1, \ldots, x_{i-1}, x_i, x_{i+1}, \ldots, x_n$, we may express any vector. (The x's form a basis). If $\mathfrak{S}'$ is linearly independent, we are done. If it is not, we apply the theorem of §5 again and again the same way until we reach a linearly independent set, containing $y_1, \ldots, y_m$, in terms of which we may express every vector in $\mathfrak{V}$. This last set is a basis containing the y's.

§7. DIMENSION OF A VECTOR SPACE

THEOREM 1. The number of elements in any basis of a finite dimensional vector space is the same as in any other basis.

PROOF. The proof of this theorem is a slight refinement of the theorem of §6 and proves, in fact, a slightly stronger assertion. Let $\mathfrak{X} = (x_1, \ldots. x_n)$ and $\mathfrak{Y} = (y_1, \ldots, y_m)$ be two finite sets of vectors each with one of the two defining properties of a basis; i.e. we assume that every vector in $\mathfrak{V}$ is a linear combination of the x's (but not that the x's are linearly independent), and we assume that the y's are linearly independent, (but not that every vector is a linear combination of them). We may apply the theorem of §5, just as above, to the set $\mathfrak{S}$:

$$y_m, x_1, x_2, \ldots, x_n.$$

Again we know that every vector is a linear combination of vectors of $\mathfrak{S}$ and that $\mathfrak{S}$ is linearly dependent. Reasoning just as before we obtain a set $\mathfrak{S}'$,

$$y_m, x_1, x_2, \ldots, x_{i-1}, x_{i+1}, \ldots, x_n,$$

again with the property that every vector is a linear combination of vectors of $\mathfrak{S}'$. Now we write y_{m-1} in front of the vectors of $\mathfrak{S}'$, and apply the same argument. Continuing in this way it is clear that the x's will not be exhausted before the y's are, since then the remaining y's would have to be linearly independent, and at the same time linear combinations, of the ones already incorporated into $\mathfrak{S}$. In other words after the argument has been applied m times we obtain a set with the same property the x's had, and this set will differ from the x's in that m of them have been replaced by y's. This seemingly innocent statement is what we are after: it implies that $n \geq m$. Consequently if $\mathfrak{X}$ and $\mathfrak{Y}$ are both bases (so that they each have both properties) then $n \geq m$ and $m \geq n$.

DEFINITION. The number of elements in a basis of a finite dimensional vector space $\mathfrak{V}$ is the linear <u>dimension</u> of $\mathfrak{V}$.

This definition (together with the fact that we have already exhibited, in §6, one particular basis in $\mathfrak{C}_n$) finally justifies our terminology and enables us to announce the pleasant result: n-dimensional Euclidean space is n-dimensional.

As a corollary of Theorem 1 we have

THEOREM 2. Any $n+1$ vectors in an n-dimensional vector space $\mathfrak{V}$ are linearly dependent. A set of n vectors in $\mathfrak{V}$ is a basis if and only if it is linearly independent, or, alternatively, if and only if every vector in $\mathfrak{V}$ is a linear combination of elements of the set.

§8. ISOMORPHISM OF VECTOR SPACES

As an application of the notion of linear basis, or coordinate system, we shall now fulfill one of our promises by showing that every finite dimensional vector space is essentially the same as some $\mathfrak{C}_n$. To make this statement precise we lay down the following definition.

DEFINITION. Two vector spaces $\mathfrak{U}$ and $\mathfrak{V}$ are <u>isomorphic</u> if there is a one to one correspondence between the vectors x of $\mathfrak{U}$ and the vectors y of $\mathfrak{V}$, say $y = T(x)$, such that $T(\alpha_1 x_1 + \alpha_2 x_2) = \alpha_1 T(x_1) + \alpha_2 T(x_2)$. In other words $\mathfrak{U}$ and $\mathfrak{V}$ are isomorphic if there is a one to one correspondence (an <u>isomorphism</u>) between their elements which preserves all linear relations.

It is easy to see that isomorphic finite dimensional vector spaces have the same dimension: to a basis in one space there corresponds a basis in the other space. Thus dimension is an isomorphism invariant: we shall now show that it is the only isomorphism invariant, i.e. that any two complex vector spaces of the same finite dimension are isomorphic. Since the isomorphism of $\mathfrak{U}$ and $\mathfrak{V}$ on the one hand, and of $\mathfrak{V}$ and $\mathfrak{W}$ on the other hand, implies that $\mathfrak{U}$ and $\mathfrak{W}$ are isomorphic, it will be sufficient to prove the following theorem.

THEOREM. Every n-dimensional complex vector space $\mathfrak{V}$ is isomorphic to $\mathfrak{C}_n$.

PROOF. Let $x_1, \ldots, x_n$ be any basis in $\mathfrak{V}$. For any x in $\mathfrak{V}$ we have $x = \xi_1 x_1 + \cdots + \xi_n x_n$, and we know that the numbers $\xi_1, \ldots, \xi_n$ are uniquely determined by x. We consider the one to one correspondence

$$x \rightleftarrows \{\xi_1, \ldots, \xi_n\}$$

between $\mathfrak{V}$ and $\mathfrak{C}_n$. If $y = \eta_1 x_1 + \cdots + \eta_n x_n$ then

$$\alpha x + \beta y = (\alpha \xi_1 + \beta\eta_1)x_1 + \cdots + (\alpha\xi_n + \beta\eta_n)x_n:$$

this establishes the desired isomorphism.

One might be tempted to say that from now on it would be silly to try to preserve an appearance of generality by talking of the general n-dimensional vector space, since we know that from the point of view of studying linear problems isomorphic vector spaces are indistinguishable, and we may as well always study $\mathfrak{C}_n$. There is one catch. The most important properties of vectors and vector spaces are those which are independent of coordinate systems, or, in other words, those which are invariant under isomorphisms. The correspondence between $\mathfrak{V}$ and $\mathfrak{C}_n$ was, however, established by choosing a coordinate system: were we always to study $\mathfrak{C}_n$ we would always be tied down to this particular coordinate system, or else we would always be faced with the problem of showing that our definitions and theorems are independent of the coordinate system in which they are stated. (This horrible dilemma will become clear later on the few occasions when we shall have to use a particular coordinate system to give a definition). Accordingly in the greater part of this book we shall ignore the theorem just proved and treat n-dimensional vector spaces as self respecting entities, independently of any basis. Besides the reasons just mentioned there is another reason for doing this: many special examples of vector spaces, such for example as $\mathfrak{P}_n$, would lose a lot of their intuitive content if we were to transform them into $\mathfrak{C}_n$ and speak only of coordinates. In studying vector spaces, such as $\mathfrak{P}$, and their relation to other vector spaces, it is important to be able to handle them with equal ease in different coordinate systems, or, and this is essentially the same thing, to be able to handle them without using any coordinate system.

§9. LINEAR MANIFOLDS

The objects of interest in geometry are not only the points of the space under consideration, but also its lines, planes, etc. The analogs, in general vector spaces, of these higher dimensional elements are introduced by the following definition.

> DEFINITION. A non empty subset $\mathfrak{M}$ of a vector space $\mathfrak{V}$ is a <u>linear manifold</u> if along with every pair, x,y, of vectors contained in $\mathfrak{M}$, every linear combination, $\alpha x + \beta y$, is also contained in $\mathfrak{M}$.

A word of warning: along with x a linear manifold also contains $x - x = 0$. Hence if we interpret linear manifolds as lines, planes, etc., we must be careful to consider only lines and planes that pass through the origin.

A linear manifold $\mathfrak{M}$ in a vector space $\mathfrak{V}$ is itself a vector space: we leave it to the reader to verify that with the same definitions of addition and scalar multiplication as we had in $\mathfrak{V}$, $\mathfrak{M}$ satisfies the axioms A and B of §1.

Two special examples of linear manifolds are, (i) the set $\mathfrak{O}$ consisting of the origin only, and (ii) the whole space $\mathfrak{V}$. Less trivial examples are the following.

(1) Let n and m be any two positive integers,

$m \leq n$. Let $\mathfrak{M}$ be the set of all vectors $x = \{\xi_1, \ldots, \xi_n\}$ in $\mathfrak{C}_n$ for which $\xi_1 = \ldots = \xi_m = 0$.

(2) With m and n as before, we consider the space $\mathfrak{P}_n$, and any m real numbers, $t_1, \ldots, t_m$. Let $\mathfrak{M}$ be the set of all vectors $x = x(t)$ in $\mathfrak{P}_n$ for which $x(t_1) = \cdots = x(t_m) = 0$.

(3) Let $\mathfrak{M}$ be the set of all vectors $x = x(t)$ in $\mathfrak{P}$ for which $x(t) \equiv x(-t)$ holds identically in t.,

We need some notation and terminology. For any collection, $\mathfrak{M}_\gamma$, of subsets of a given set, say, for example, for a collection of linear manifolds in a vector space $\mathfrak{V}$, we write $\bigcap_\gamma \mathfrak{M}_\gamma$ for the intersection of all $\mathfrak{M}_\gamma$, i.e. the set of points common to all of them. Also if $\mathfrak{M}$ and $\mathfrak{N}$ are subsets of a set we write $\mathfrak{M} \subset \mathfrak{N}$ if $\mathfrak{M}$ is a subset of $\mathfrak{N}$, i.e. if every element of $\mathfrak{M}$ lies also in $\mathfrak{N}$. (Observe that we do not exclude the possibility $\mathfrak{M} = \mathfrak{N}$; thus we write $\mathfrak{V} \subset \mathfrak{V}$ as well as $\mathfrak{O} \subset \mathfrak{V}$). For a finite collection, $\mathfrak{M}_1, \ldots, \mathfrak{M}_n$, we shall write $\mathfrak{M}_1 \cap \ldots \cap \mathfrak{M}_n$ in place of $\bigcap_j \mathfrak{M}_j$; in case two linear manifolds, $\mathfrak{M}$ and $\mathfrak{N}$, are such that $\mathfrak{M} \cap \mathfrak{N} = \mathfrak{O}$ we shall say that $\mathfrak{M}$ and $\mathfrak{N}$ are <u>disjoint</u>.

§10. CALCULUS OF LINEAR MANIFOLDS

THEOREM 1. The intersection of any collection of linear manifolds is a linear manifold.

PROOF. If we use an index γ to tell apart the members of the collection, so that the given linear manifolds are $\mathfrak{M}_\gamma$, let us write

$$\mathfrak{M} = \bigcap_\gamma \mathfrak{M}_\gamma .$$

Since every $\mathfrak{M}_\gamma$ contains 0, so does $\mathfrak{M}$, hence $\mathfrak{M}$ is non empty. If x and y belong to $\mathfrak{M}$ (i.e. to all $\mathfrak{M}_\gamma$), then $\alpha x + \beta y$ also belongs to all $\mathfrak{M}_\gamma$ --

hence $\mathfrak{M}$ is a linear manifold.

To give an application of this theorem, suppose that $\mathfrak{S}$ is an arbitrary set of vectors (not necessarily a linear manifold) in a vector space $\mathfrak{V}$. There certainly exist linear manifolds $\mathfrak{M}$ containing every element of $\mathfrak{S}$, $\mathfrak{S} \subset \mathfrak{M}$ ---- the whole space $\mathfrak{V}$ is, for example, such a linear manifold. Let us denote by $\mathfrak{M}$ the intersection of all linear manifolds containing $\mathfrak{S}$: it is clear that $\mathfrak{M}$ is itself a linear manifold containing $\mathfrak{S}$. $\mathfrak{M}$ is, moreover, the smallest such manifold: if $\mathfrak{S}$ is also contained in the linear manifold $\mathfrak{N}$, $\mathfrak{S} \subset \mathfrak{N}$, then $\mathfrak{M} \subset \mathfrak{N}$. The manifold $\mathfrak{M}$ so defined is called the linear manifold <u>spanned</u> by $\mathfrak{S}$. The connection between the notion of spanning and the concepts studied in §§4-8 is the following.

> THEOREM 2. If $\mathfrak{S}$ is any set of vectors in a vector space $\mathfrak{V}$, and $\mathfrak{M}$ is the linear manifold spanned by $\mathfrak{S}$, then $\mathfrak{M}$ is the same as the set of all linear combinations of elements of $\mathfrak{S}$.

PROOF. It is clear that a linear combination of linear combinations of elements of $\mathfrak{S}$ may again be written as a linear combination of elements of $\mathfrak{S}$. Hence the set of all linear combinations of elements of $\mathfrak{S}$ is a linear manifold containing $\mathfrak{S}$: it follows that this manifold must also contain $\mathfrak{M}$. Now turn the argument around: $\mathfrak{M}$ contains $\mathfrak{S}$ and is a linear manifold, hence $\mathfrak{M}$ contains all linear combinations of elements of $\mathfrak{S}$

We see therefore that in our new terminology we may define a linear basis as a set of linearly independent vectors which spans the whole space.

As an easy consequence, whose proof we leave to the reader, of Theorem 2 we have:

THEOREM 3. If $\mathfrak{H}$ and $\mathfrak{K}$ are any two linear manifolds and $\mathfrak{M}$ is the linear manifold spanned by $\mathfrak{H}$ and $\mathfrak{K}$ together, then $\mathfrak{M}$ is the same as the set of all vectors of the form x+y, with x in $\mathfrak{H}$ and y in $\mathfrak{K}$.

Prompted by this theorem we shall use the notation $\mathfrak{H} + \mathfrak{K}$ for the linear manifold $\mathfrak{M}$ spanned by $\mathfrak{H}$ and $\mathfrak{K}$.

§11. DIMENSION OF A LINEAR MANIFOLD

THEOREM 1. A linear manifold $\mathfrak{M}$ in an n-dimensional vector space $\mathfrak{V}$ is a vector space of dimension $\leq$ n.

PROOF. It is possible to give a deceptively short proof of this theorem, that runs as follows. Every set of n+1 vectors in $\mathfrak{V}$ is linearly dependent, hence the same is true of $\mathfrak{M}$; hence in particular the number of elements in any basis of $\mathfrak{M}$ is $\leq$ n; Q.E.D.

The trouble is that we defined dimension n by requiring in the first place that there exist a finite basis, and then demanding that this basis contain exactly n elements. The proof above shows only that no basis can contain more than n elements; it does not show that any basis exists. Once the difficulty is observed, however, it is easy to fill the gap. If $\mathfrak{M} = \mathfrak{O}$ then $\mathfrak{M}$ is 0-dimensional and we're done. If $\mathfrak{M}$ contains a vector $x_1 \neq 0$, let $\mathfrak{M}_1 \subset \mathfrak{M}$ be the linear manifold spanned by x_1. If $\mathfrak{M}_1 = \mathfrak{M}$, $\mathfrak{M}$ is 1-dimensional, and we are done. If $\mathfrak{M}_1 \neq \mathfrak{M}$, let x_2 be an element of $\mathfrak{M}$ not contained in $\mathfrak{M}_1$, and let $\mathfrak{M}_2$ be the linear manifold spanned by x_1 and x_2; and so on. Now we may legitimately employ the argument given above: after no more than n steps of this sort the process reaches an end, since (by Theorem 2, §7) we cannot find $n+1$ linearly independent vectors.

An important consequence of this second and correct proof of Theorem 1 is the following.

THEOREM 2. Given any linear manifold $\mathfrak{M}$ in an n-dimensional vector space $\mathfrak{V}$ we may find a basis $x_1, \ldots, x_m, x_{m+1} \ldots, x_n$ in $\mathfrak{V}$, so that $x_1, \ldots, x_m$ are in $\mathfrak{M}$ and form, therefore, a basis in $\mathfrak{M}$.

It is easy to manufacture examples illustrating the concepts and results of the last two sections. One example is this: a polynomial (for us a vector $x = x(t)$ in $\mathfrak{P}$) is called <u>even</u> if $x(-t) \equiv x(t)$, (see §9,(3)) and it is called <u>odd</u> if $x(-t) \equiv -x(t)$. Both the class of even polynomials and the class of odd polynomials are linear manifolds in $\mathfrak{P}$; these two linear manifolds, say $\mathfrak{M}$ and $\mathfrak{N}$, are disjoint, and $\mathfrak{M} + \mathfrak{N} = \mathfrak{P}$. (Proof?)

§12. CONJUGATE SPACES

DEFINITION. A <u>linear functional</u>, $y = y(x)$, on a vector space $\mathfrak{V}$, is a scalar valued function defined for every vector x, with the property that (identically in the vectors x_1 and x_2 and scalars α_1 and α_2)

$$y(\alpha_1 x_1 + \alpha_2 x_2) = \alpha_1 y(x_1) + \alpha_2 y(x_2).$$

Let us look at some examples of linear functionals.

(1) For $x = \{\xi_1, \ldots, \xi_n\}$ in $\mathfrak{C}_n$, define $y(x) = \xi_1$. More generally, let $\alpha_1, \ldots, \alpha_n$ be any n complex numbers, and define, for

$x = \{\xi_1, \ldots \xi_n\}$, $y(x) = \alpha_1 \xi_1 + \ldots + \alpha_n \xi_n$.

We observe that in any vector space and for any

linear functional y,

$$y(0) = y(0 \cdot 0) = 0 \cdot y(0) = 0;$$

hence a linear functional as we defined it is sometimes called homogeneous. In particular in $\mathfrak{C}_n$

$$y(x) = \alpha_1 \xi_1 + \cdots + \alpha_n \xi_n + \beta$$

is not a linear functional, unless $\beta = 0$.

(2) For $x = x(t)$ in $\mathfrak{P}$, define $y(x) = x(0)$. More generally for any n real numbers $t_1, \ldots, t_n$ and any n complex numbers $\alpha_1, \ldots, \alpha_n$, define

$$y(x) = \alpha_1 x(t_1) + \cdots + \alpha_n x(t_n).$$

Another example, in a sense a limiting case of the one just given, is the following. Let (a,b) be any interval on the t axis, and let $\alpha(t)$ be any complex valued integrable function defined for t in (a,b). We write for every $x = x(t)$ in $\mathfrak{P}$,

$$y(x) = \int_a^b \alpha(t)x(t)\,dt.$$

(3) In an arbitrary vector space $\mathfrak{V}$, define $y(x) = 0$ for every x in $\mathfrak{V}$.

This last example is a first hint of a general situation. Let $\mathfrak{V}$ be any vector space and let $\mathfrak{V}'$ be the collection of all linear functionals on $\mathfrak{V}$. Let us denote by 0 the linear functional defined in (3) (compare the comment at the end of §3); if $y_1 = y_1(x)$ and $y_2 = y_2(x)$, and if α_1 and α_2 are any scalars, let us denote by y_3 the expression

$$y_3 = y_3(x) = \alpha_1 y_1(x) + \alpha_2 y_2(x).$$

It is easy to see that y_3 is a linear functional; we denote it by $\alpha_1 y_1 + \alpha_2 y_2$.

With these definitions of the linear concepts, (zero, addition, scalar multiplication), $\mathfrak{V}'$ forms a vector space, the conjugate space of $\mathfrak{V}$.

§13. NOTATION FOR LINEAR FUNCTIONALS

Before studying linear functionals and conjugate spaces in more detail we wish to introduce a notation that may appear weird at first sight but which will clarify many situations later on. We have already used the trick of denoting such a composite object as a linear functional, $y(x)$, by a single letter y. Sometimes, however, it is necessary to use the function notation and indicate somehow that if y is a linear functional on $\mathfrak{V}$ and if x is a vector in $\mathfrak{V}$, then $y(x)$ is a fixed scalar. The notation we propose to adopt here, instead of writing y followed by x in parentheses, is x and y enclosed between square brackets and separated by a comma: in other words we shall write $[x,y]$ in place of $y(x)$. Because of the unusual nature of this notation we shall expend on it some further verbiage.

As we have just pointed out $[x,y]$ is a substitute for the function notation $y(x)$; $[x,y]$ is the scalar value we obtain if we take the value of the linear functional y at the vector x. Let us take an analogous situation (but not concerned with <u>linear</u> functionals). Let x be a real variable; let $y = y(x)$ be the function $y(x) = x^2$. The notation $[x,y]$ is a symbolic way of writing down the recipe for the actual operations performed; it corresponds to the sentence [take a number x, and square it].

Using this notation we may sum up: to every vector space $\mathfrak{V}$ we make correspond the conjugate space $\mathfrak{V}'$ consisting of all linear functionals on $\mathfrak{V}$; to every pair x and y, where x is a vector in $\mathfrak{V}$ and y is a linear functional in $\mathfrak{V}'$, we make correspond the scalar $[x,y]$ defined to be the value of y at x. In terms of the symbol $[x,y]$ the defining property of a linear functional is

$$(1) \quad [\alpha_1 x_1 + \alpha_2 x_2, y] = \alpha_1 [x_1, y] + \alpha_2 [x_2, y],$$

and the definition of the linear operations on linear functionals is

(2) $[x, \alpha_1 y_1 + \alpha_2 y_2] = \alpha_1 [x,y_1] + \alpha_2 [x,y_2]$.

The two relations together are expressed by saying that $[x,y]$ is a bilinear functional of the vectors x in $\mathfrak{V}$ and y in $\mathfrak{V}'$.

§14. BASES IN CONJUGATE SPACES

One more word before embarking on the proofs of the important theorems. The concept of conjugate space was defined without any reference to coordinate systems; a glance at the following proofs will show a superabundance of coordinate systems. We wish to point out that this phenomenon is inevitable in this case: we shall be establishing results concerning dimension, and dimension is the one concept (so far) whose very definition is given in terms of a basis.

THEOREM 1. If $\mathfrak{V}$ is an n-dimensional vector space, if $\mathfrak{X} = (x_1, \ldots, x_n)$ is a basis in $\mathfrak{V}$, and if $\alpha_1, \ldots, \alpha_n$ is any set of n scalars, then there is one and only one linear functional y on $\mathfrak{V}$ such that $[x_i, y] = \alpha_i$ for $i = 1, \ldots, n$.

PROOF. Every x in $\mathfrak{V}$ may be written in the form $x = \xi_1 x_1 + \cdots + \xi_n x_n$ in one and only one way; if y is any linear functional then

$$[x,y] = \xi_1 [x_1,y] + \cdots + \xi_n [x_n,y].$$

From this relation the uniqueness of y is clear: if $[x_i,y] = \alpha_i$, then the value of $[x,y]$ is determined for every x by $[x,y] = \sum_i \xi_i \alpha_i$. The argument can also be turned around: if we define

$$[x,y] = \xi_1 \alpha_1 + \cdots + \xi_n \alpha_n$$

then y is indeed a linear functional and $[x_i,y] = \alpha_i$.

THEOREM 2. If $\mathfrak{V}$ is an n-dimensional vector space and $\mathfrak{X} = (x_1, \ldots, x_n)$ is a basis in $\mathfrak{V}$, then there is a uniquely determined basis $\mathfrak{X}'$ in $\mathfrak{V}'$, $\mathfrak{X}' = (y_1, \ldots, y_n)$, called the dual basis of $\mathfrak{X}$, with the property that $[x_i, y_j] = \delta_{ij}$. Consequently the conjugate space of an n-dimensional vector space is n-dimensional.

PROOF. It follows from Theorem 1 that for each $j = 1, \ldots, n$ a unique y_j in $\mathfrak{V}'$ can be found for which $[x_i, y_j] = \delta_{ij}$; we have only to prove that the set $\mathfrak{X}' = (y_1, \ldots, y_n)$ is a basis in $\mathfrak{V}'$.

In the first place $\mathfrak{X}'$ is a linearly independent set, for if we had $\alpha_1 y_1 + \cdots + \alpha_n y_n = 0$, in other words if

$$[x, \alpha_1 y_1 + \cdots + \alpha_n y_n] = \alpha_1 [x, y_1] + \cdots + \alpha_n [x, y_n] = 0$$

for all x, then we should have, for $x = x_i$,

$$0 = \sum_j \alpha_j [x_i, y_j] = \sum_j \alpha_j \delta_{ij} = \alpha_i.$$

In the second place every y in $\mathfrak{V}'$ is a linear combination of $y_1, \ldots, y_n$. To prove this write $[x_i, y] = \alpha_i$; then for $x = \sum_i \xi_i x_i$ we have

(1) $$[x, y] = \xi_1 \alpha_1 + \cdots + \xi_n \alpha_n.$$

On the other hand

(2) $$[x, y_j] = \sum_i \xi_i [x_i y_j] = \xi_j$$

so that, substituting in (1),

$$[x, y] = \alpha_1 [x, y_1] + \cdots + \alpha_n [x, y_n]$$
$$= [x, \alpha_1 y_1 + \cdots + \alpha_n y_n].$$

Consequently $y = \alpha_1 y_1 + \cdots + \alpha_n y_n$, and the proof of the theorem is complete.

We shall need also the following easy consequence of Theorem 2.

THEOREM 3. If x' and x'' are any two different vectors of the n-dimensional vector space $\mathfrak{V}$, then there exists a linear functional y on $\mathfrak{V}$ such that $[x',y] \neq [x'',y]$; or, equivalently, to any non zero vector x in $\mathfrak{V}$ there corresponds a y in $\mathfrak{V}'$ such that $[x,y] \neq 0$.

PROOF. That the two statements in the theorem are indeed equivalent is seen by considering $x = x' - x''$. We shall, accordingly, prove only the latter statement.

Let $\mathfrak{X} = (x_1, \ldots, x_n)$ be any basis in $\mathfrak{V}$, and let $\mathfrak{X}' = (y_1, \ldots, y_n)$ be the dual basis in $\mathfrak{V}'$. If $x = \sum_i \xi_i x_i$, then (see (2)) $[x,y_j] = \xi_j$. Hence if $[x,y] = 0$ for all y, in particular if $[x,y_j] = 0$ for $j = 1, \ldots, n$, then $x = 0$.

§15. REFLEXIVITY OF FINITE DIMENSIONAL SPACES

It is natural to think that if the conjugate space, $\mathfrak{V}'$, of a vector space $\mathfrak{V}$, and the relations between a space and its conjugate space, are of any interest at all for $\mathfrak{V}$, then they are of just as much interest for $\mathfrak{V}'$. In other words we may form the conjugate space $(\mathfrak{V}')'$ of $\mathfrak{V}'$: for simplicity of notation we shall denote it by $\mathfrak{V}''$. The verbal description of an element of $\mathfrak{V}''$ is clumsy: such an element is a linear functional of linear functionals. It is, however, at this point that the greatest advantage of the notation $[x,y]$ appears: by means of it, it is easy to discuss $\mathfrak{V}$ and its relation to $\mathfrak{V}''$.

If we consider the symbol $[x,y]$ for some fixed $y = y_0$, we obtain nothing new: $[x,y_0]$ is merely another way of writing the function $y_0 = y_0(x)$. If, however, we consider the symbol $[x,y]$ for some fixed $x = x_0$, then we observe that the function $[x_0,y]$ of the vectors

y in $\mathfrak{B}'$ is a scalar valued function which is linear (see §13, (2)): in other words $[x_0, y]$ is a linear functional on $\mathfrak{B}'$ and, as such, an element of $\mathfrak{B}''$.

By this method we have exhibited <u>some</u> linear functionals on $\mathfrak{B}$: have we exhibited them all? For the finite dimensional case the following theorem furnishes the affirmative answer.

THEOREM. If $\mathfrak{B}$ is a finite dimensional vector space then corresponding to every linear functional $z_0 = z_0(y)$ on $\mathfrak{B}'$ there is a vector x_0 in $\mathfrak{B}$ such that $z_0(y) = [x_0, y] = y(x_0)$; the correspondence $z_0 \rightleftarrows x_0$, the so-called <u>natural correspondence</u> between $\mathfrak{B}''$ and $\mathfrak{B}$, is an isomorphism.

PROOF. Let us view the correspondence from the standpoint of going from $\mathfrak{B}$ to $\mathfrak{B}''$: in other words to every x_0 in $\mathfrak{B}$ we make correspond a vector z_0 in $\mathfrak{B}''$ defined by $z_0(y) = y(x_0)$ for every y in $\mathfrak{B}'$. Since $[x, y] = y(x)$ is linear in x, the transformation $x_0 \to z_0$ is linear.

We shall show that this transformation is one to one, as far as it goes: in other words if to x' in $\mathfrak{B}$ there corresponds the linear functional $z' = z'(y) = [x', y]$ on $\mathfrak{B}''$, and to x'' there corresponds $z'' = z''(y) = [x'', y]$, and if $z' = z''$ then $x' = x''$. To say that $z' = z''$ means that $[x', y] = [x'', y]$ for every y in $\mathfrak{B}'$: it follows from Theorem 3 of §14 that $x' = x''$.

The last two paragraphs together show that the set of those linear functionals $z = z(y)$ on $\mathfrak{B}'$ (i.e. elements of $\mathfrak{B}''$) which do have the desired form, $z(y) = [x, y]$ for a suitable x in $\mathfrak{B}$, form a linear manifold in $\mathfrak{B}''$ which is isomorphic to $\mathfrak{B}$ and which is therefore n-dimensional. But the n-dimensionality of

$\mathfrak{V}$ implies that of $\mathfrak{V}'$, which in turn implies that $\mathfrak{V}''$ is n-dimensional. It follows that $\mathfrak{V}''$ must coincide with the n-dimensional linear manifold just described, and the proof of the theorem is complete.

It is important to observe that this theorem shows not only that $\mathfrak{V}$ and $\mathfrak{V}''$ are isomorphic -- this much is trivial from the fact that they have the same dimension -- but that the natural correspondence is an isomorphism. This property of vector spaces is called <u>reflexivity</u>: every finite dimensional vector space is reflexive.

It is extremely convenient to be mildly sloppy about $\mathfrak{V}''$: for finite dimensional vector spaces we shall identify $\mathfrak{V}''$ with $\mathfrak{V}$ (by the natural isomorphism) and we shall say that the element $z_0(y) = [x_0, y]$ of $\mathfrak{V}''$ is the <u>same</u> as the element x_0. In this language it is very easy to express the relation between a basis $\mathfrak{X}$, in $\mathfrak{V}$, and the dual basis of its dual basis, in $\mathfrak{V}''$: the symmetry of the relation $[x_i, y_j] = \delta_{ij}$ shows that $\mathfrak{X}'' = \mathfrak{X}$.

§16. ANNIHILATORS OF LINEAR MANIFOLDS

DEFINITION. The <u>annihilator</u>, $\mathfrak{S}^0$, of any subset $\mathfrak{S}$ of a vector space $\mathfrak{V}$, ($\mathfrak{S}$ need not be a linear manifold), is the set of all vectors y in $\mathfrak{V}'$ for which $[x, y]$ is identically zero for all x in $\mathfrak{S}$.

Thus $\mathfrak{O}^0 = \mathfrak{V}'$ and $\mathfrak{V}^0 = \mathfrak{O}$ ($\subset \mathfrak{V}'$). If $\mathfrak{V}$ is finite dimensional and $\mathfrak{S}$ contains a non zero vector, i.e. $\mathfrak{S} \neq \mathfrak{O}$, then Theorem 3 of §14 shows that $\mathfrak{S}^0 \neq \mathfrak{V}'$.

THEOREM 1. If $\mathfrak{M}$ is an m-dimensional linear manifold in an n-dimensional vector space $\mathfrak{V}$, then $\mathfrak{M}^0$ is an n-m dimensional linear manifold in $\mathfrak{V}'$.

PROOF. We leave it to the reader to verify that $\mathfrak{M}^{\circ}$ (in fact $\mathfrak{S}^{\circ}$, for an arbitrary $\mathfrak{S}$) is always a linear manifold; we shall prove only the statement concerning the dimension of $\mathfrak{M}^{\circ}$.

Let $\mathfrak{X} = (x_1, \ldots, x_n)$ be a basis in $\mathfrak{V}$ whose first m elements are in $\mathfrak{M}$ (and are therefore a basis for $\mathfrak{M}$); let $\mathfrak{X}' = (y_1, \ldots, y_n)$ be the dual basis in $\mathfrak{V}'$. We denote by $\tilde{\mathfrak{M}}$ the linear manifold (in $\mathfrak{V}'$) spanned by $y_{m+1}, \ldots, y_n$; clearly $\tilde{\mathfrak{M}}$ has dimension $n-m$. We shall prove that $\mathfrak{M}^{\circ} = \tilde{\mathfrak{M}}$.

If x is any vector in $\mathfrak{M}$, then x is a linear combination of $x_1, \ldots, x_m$, $x = \sum_{i=1}^{m} \xi_i x_i$, and for any $j = m+1, \ldots, n$ we have

$$[x,y_j] = \sum_{i=1}^{m} \xi_i [x_i,y_j] = 0.$$

In other words for $j = m+1, \ldots, n$, y_j is in $\mathfrak{M}^{\circ}$; it follows that $\tilde{\mathfrak{M}}$ is contained in $\mathfrak{M}^{\circ}$, $\tilde{\mathfrak{M}} \subset \mathfrak{M}^{\circ}$. Suppose on the other hand that y is any element of $\mathfrak{M}^{\circ}$. Since y, being in $\mathfrak{V}'$, is a linear combination of the basis vectors $y_1, \ldots, y_n$, we may write $y = \sum_{j=1}^{n} \eta_j y_j$. Since, by hypothesis, y is in $\mathfrak{M}^{\circ}$ we have for every $i = 1, \ldots, m$

$$0 = [x_i,y] = \sum_{i=1}^{n} \eta_j [x_i,y_j] = \eta_i;$$

in other words y is a linear combination of $y_{m+1}, \ldots, y_n$. This proves that y is in $\tilde{\mathfrak{M}}$, and consequently that $\mathfrak{M}^{\circ} \subset \tilde{\mathfrak{M}}$, and the theorem follows.

THEOREM 2. If $\mathfrak{M}$ is a linear manifold in a finite dimensional vector space $\mathfrak{V}$ then $\mathfrak{M}^{\circ\circ} (= (\mathfrak{M}^{\circ})^{\circ}) = \mathfrak{M}$.

PROOF. Observe that we use here the convention, established at the end of §15, identifying $\mathfrak{V}$ and $\mathfrak{V}''$. By definition $\mathfrak{M}^{\circ\circ}$ is the set of all vectors x for which y in $\mathfrak{M}^{\circ}$ implies $[x,y] = 0$. Since, by definition of $\mathfrak{M}^{\circ}$, $[x,y] = 0$ for all x in $\mathfrak{M}$ and all y in

$\mathfrak{M}^0$, it follows that $\mathfrak{M} \subset \mathfrak{M}^{00}$. The desired conclusion now follows from a dimension argument. Let $\mathfrak{M}$ be m-dimensional; then the dimension of $\mathfrak{M}^0$ is n-m, and that of $\mathfrak{M}^{00}$ is n-(n-m) = m. Hence $\mathfrak{M} = \mathfrak{M}^{00}$, as was to be proved.

To see an example of an annihilator the reader might determine what the most general linear functional in $\mathfrak{C}$ looks like, (see §12, (1)) and then find the annihilator of the manifold described in §9,(1). Also: it is an easy consequence of Theorem 2 of this section that for any subset $\mathfrak{S}$ of a finite dimensional vector space, $\mathfrak{S}^{00}$ is the same as the linear manifold spanned by $\mathfrak{S}$. Proof?

§17. DIRECT SUMS

In this section we shall describe a general method of making new vector spaces out of old ones.

DEFINITION. If $\mathfrak{U}$ and $\mathfrak{V}$ are any two vector spaces their <u>direct sum</u>, $\mathfrak{W} = \mathfrak{U} \oplus \mathfrak{V}$ is the set of all pairs $\{x,y\}$ with x in $\mathfrak{U}$ and y in $\mathfrak{V}$, and with the linear operations defined by the formula

$$\alpha_1 \{x_1, y_1\} + \alpha_2 \{x_2, y_2\} = \{\alpha_1 x_1 + \alpha_2 x_2, \alpha_1 y_1 + \alpha_2 y_2\}$$

We observe that the formation of the direct sum is analogous to the way in which the Euclidean plane is constructed from the x and y axes.

We investigate the relation of this notion to some of our earlier ones.

The set of all vectors (in $\mathfrak{W}$) of the form $\{x,0\}$ is a linear manifold in $\mathfrak{W}$; the correspondence $\{x,0\} \rightleftarrows x$ shows that this linear manifold is isomorphic to $\mathfrak{U}$. It is convenient, once more, to indulge in a logical inaccuracy and to identify x and $\{x,0\}$ and speak of $\mathfrak{U}$ as a linear manifold in $\mathfrak{W}$. Similarly, of course, the vectors y of $\mathfrak{V}$ may be identified with the vectors of the form $\{0,y\}$ in $\mathfrak{W}$, and we may consider $\mathfrak{V}$ as a linear manifold in $\mathfrak{W}$. This terminology is not, to be sure, quite exact, but the logical difficulty is much easier to get around here than it was in the case of the second conjugate space. We could have defined the direct sum of $\mathfrak{U}$ and $\mathfrak{V}$ to be the set consisting of all x's in $\mathfrak{U}$, all y's in $\mathfrak{V}$, and all those pairs $\{x,y\}$ for which $x \neq 0$ and $y \neq 0$. This definition yields a theory analogous in every detail to the one we shall develop, but it makes it a nuisance to prove theorems because of the case distinctions it necessitates. It is clear, however, that from the point of view of this definition $\mathfrak{U}$ is actually a subset of $\mathfrak{U} \oplus \mathfrak{V}$. In this sense then, or in the isomorphism sense of the definition we did adopt, we raise the question: what is the relation between $\mathfrak{U}$ and $\mathfrak{V}$ when we consider these spaces as subspaces of the big space $\mathfrak{W}$?

THEOREM. If $\mathfrak{U}$ and $\mathfrak{V}$ are linear manifolds in a vector space $\mathfrak{W}$, then the following three conditions are equivalent.

(1) $\mathfrak{W} = \mathfrak{U} \oplus \mathfrak{V}$

(2) $\mathfrak{U}$ and $\mathfrak{V}$ are disjoint and $\mathfrak{W} = \mathfrak{U} + \mathfrak{V}$, (§10);

(3) Every vector z in $\mathfrak{W}$ may be written in the form $z = x+y$, with x in $\mathfrak{U}$ and y in $\mathfrak{V}$, in one and only one way.

PROOF. We shall prove the implications (1) $\Longrightarrow$ (2) $\Longrightarrow$ (3) $\Longrightarrow$ (1).

(1) $\Longrightarrow$ (2). We assume that $\mathfrak{W} = \mathfrak{U} \oplus \mathfrak{V}$. If $z = \{x,y\}$ lies in both $\mathfrak{U}$ and $\mathfrak{V}$ then $x = y = 0$, so that $z = 0$, and $\mathfrak{U}$ and $\mathfrak{V}$ are disjoint. Since the representation $z = \{x,0\} + \{0,y\} = x+y$ is valid for every z. it follows also that $\mathfrak{U} + \mathfrak{V} = \mathfrak{W}$.

(2) $\Longrightarrow$ (3). If we assume (2), so that in particular $\mathfrak{U} + \mathfrak{V} = \mathfrak{W}$, then it is clear that every z in $\mathfrak{W}$ has the desired representation, $z = x+y$. To prove uniqueness we suppose that $z = x'+y'$; x and x' are in $\mathfrak{U}$ and y and y' are in $\mathfrak{V}$. It follows from $x+y = x'+y'$ that $x - x' = y' - y$. Since the left member of this last equation is in $\mathfrak{U}$ and the right member is in $\mathfrak{V}$, the disjointness of $\mathfrak{U}$ and $\mathfrak{V}$ implies that $x = x'$ and $y = y'$.

(3) $\Longrightarrow$ (1). This implication is practically indistinguishable from the definition of direct sum. If we form the direct sum, $\mathfrak{U} \oplus \mathfrak{V}$, and then identify $\{x,0\}$ and $\{0,y\}$ with x and y respectively, we are committed to identifying the sum $\{x,y\} = \{x,0\} + \{0,y\}$ with what we are assuming to be the general element $z = x+y$ of $\mathfrak{W}$; from the hypothesis that the $z = x+y$ representation is unique we conclude that the correspondence between $\{x,0\}$ and x (and also between $\{0,y\}$ and y) is one to one.

§18. DIMENSION OF A DIRECT SUM

What can be said about the dimension of a direct sum? If $\mathfrak{U}$ is n-dimensional, $\mathfrak{V}$ is m-dimensional, and $\mathfrak{W} = \mathfrak{U} \oplus \mathfrak{V}$, what is the dimension of $\mathfrak{W}$? This question is easy to answer.

THEOREM 1. The dimension of a direct sum is the sum of dimensions of its summands.

PROOF. We assert that if $x_1, \ldots, x_n$ is a basis in $\mathfrak{U}$, and if $y_1, \ldots, y_m$ is a basis in $\mathfrak{V}$, then the set $x_1, \ldots, x_n, y_1, \ldots, y_m$ (or, more precisely, the set $\{x_1,0\}, \ldots, \{x_n,0\}, \{0,y_1\}, \ldots, \{0,y_m\}$) is a basis in $\mathfrak{W}$. The easiest proof of this assertion is to use the implication (1) $\Longrightarrow$ (3) from the theorem of the preceding section. Since every z in $\mathfrak{M}$ may be written in the form $z = x+y$, where x is a linear combination of $x_1, \ldots, x_n$, and y is a linear combination of $y_1, \ldots, y_m$, it follows that our set does indeed span $\mathfrak{M}$. To show that the set is also linearly independent, suppose that

$$\alpha_1 x_1 + \cdots + \alpha_n x_n + \beta_1 y_1 + \cdots + \beta_m y_m = 0.$$

The uniqueness of the $z = x+y$ representation implies that

$$\alpha_1 x_1 + \cdots + \alpha_n x_n = \beta_1 y_1 + \cdots + \beta_m y_m = 0,$$

and hence the linear independence of the x's and of the y's implies that

$$\alpha_1 = \ldots = \alpha_n = \beta_1 = \ldots = \beta_m = 0.$$

THEOREM 2. If $\mathfrak{M}$ is any $n+m$ dimensional vector space, and $\mathfrak{U}$ is any n-dimensional linear manifold in $\mathfrak{M}$, there exists an m-dimensional linear manifold $\mathfrak{V}$ in $\mathfrak{M}$ such that $\mathfrak{M} = \mathfrak{U} \oplus \mathfrak{V}$.

PROOF. Let $x_1, \ldots, x_n$ be any basis in $\mathfrak{U}$; by the theorem of §9 we may find a set $y_1, \ldots, y_m$ of vectors in $\mathfrak{M}$ with the property that $x_1, \ldots, x_n, y_1, \ldots, y_m$ is a basis in $\mathfrak{M}$. Let $\mathfrak{V}$ be the linear manifold spanned by $y_1, \ldots, y_m$: we omit the verification that $\mathfrak{M} = \mathfrak{U} \oplus \mathfrak{V}$.

We observe that there is no reason to expect $\mathfrak{V}$ to be uniquely determined: it is easy to construct examples to show that indeed it is not. We consider a few examples of direct sums.

(1) Let $\mathfrak{U}$ be the linear manifold of all vectors $\{\xi_1, \ldots, \xi_n, \xi_{n+1}, \ldots, \xi_{n+m}\}$ in $\mathfrak{C}_{n+m}$ for which $\xi_{n+1} = \cdots = \xi_{n+m} = 0$; let $\mathfrak{V}$ be the manifold for which $\xi_1 = \cdots = \xi_n = 0$.

(2) For a fancier example we consider $\mathfrak{P}$, and we let $\mathfrak{U}$ be the set of even polynomials, and $\mathfrak{V}$ the set of odd polynomials. (See §11). The relation

$$x(t) = \frac{1}{2}\{x(t)+x(-t)\} + \frac{1}{2}\{x(t)-x(-t)\},$$

valid for every polynomial $x(t)$, shows that $\mathfrak{U} \oplus \mathfrak{V} = \mathfrak{P}$.

(3) Let $\mathfrak{U}$ be the set of vectors $\{\xi_1, \xi_2\}$ in $\mathfrak{C}_2$ for which $\xi_2 = 0$; let $\mathfrak{V}$ be the vectors for which $\xi_1 = 0$ (see (1) above); and let $\tilde{\mathfrak{V}}$ be the vectors for which $\xi_1 = \xi_2$. Then $\mathfrak{C}_2 = \mathfrak{U} \oplus \mathfrak{V} = \mathfrak{U} \oplus \tilde{\mathfrak{V}}$. Picture?

§19. CONJUGATE SPACES OF DIRECT SUMS

In most of what follows we shall view the notion of direct sum as defined for linear manifolds in a vector space $\mathfrak{V}$; this avoids fussing with the identification convention of §17 and turns out, incidentally, to be the more useful concept for our later work. We conclude, for the present, our study of direct sums, by observing the simple relation connecting conjugate spaces, annihilators, and direct sums. To emphasize our present view of direct summation we return to the letters of our earlier notation.

THEOREM. If $\mathfrak{M}$ and $\mathfrak{N}$ are linear manifolds in a vector space $\mathfrak{V}$, and if $\mathfrak{V} = \mathfrak{M} \oplus \mathfrak{N}$ then $\mathfrak{M}'$ is isomorphic to $\mathfrak{N}^o$ and $\mathfrak{N}'$ to $\mathfrak{M}^o$, and $\mathfrak{V}' = \mathfrak{M}^o \oplus \mathfrak{N}^o$

PROOF. To simplify the notation we shall use, throughout this proof, x, x', and x^o for elements of $\mathfrak{M}$, $\mathfrak{M}'$, and $\mathfrak{M}^o$, respectively, and we reserve, similarly, the letters y for $\mathfrak{M}$ and z for $\mathfrak{V}$. (This notation is not meant to suggest that there is any particular relation between, say, the vectors x in $\mathfrak{M}$

and the vectors x' in $\mathfrak{M}'$.)

If z' belongs to both $\mathfrak{M}^0$ and $\mathfrak{N}^0$, i.e. if $z'(x) \equiv z'(y) \equiv 0$, then $z'(z) = z'(x+y) = 0$, so that $\mathfrak{M}^0$ and $\mathfrak{N}^0$ are disjoint. If, moreover, z' is any vector in $\mathfrak{V}'$, we define for every $z = x+y$, $x^0(z) = z'(y)$ and $y^0(z) = z'(x)$. It is easy to see that x^0 and y^0 are linear functionals on $\mathfrak{V}$ (i.e. elements of $\mathfrak{V}'$) belonging to $\mathfrak{M}^0$ and $\mathfrak{N}^0$ respectively; since $z' = x^0+y^0$, it follows that $\mathfrak{V}'$ is indeed the direct sum of $\mathfrak{M}^0$ and $\mathfrak{N}^0$.

To establish the asserted isomorphisms, we make correspond to every x^0 a y' in $\mathfrak{N}'$ defined by $y'(y) = x^0(y)$. We leave to the reader the routine verification that the correspondence $x^0 \longrightarrow y'$ is linear and one to one, and therefore an isomorphism, between $\mathfrak{M}^0$ and $\mathfrak{N}'$; the corresponding result for $\mathfrak{N}^0$ and $\mathfrak{M}'$ follows from symmetry by interchanging x and y.

We remark, concerning our entire presentation of the theory of direct sums, that there is nothing magic about the number two: we could have defined the direct sum of any finite number of vector spaces, and proved the obvious analogs of all theorems of the last three sections, with only the notation becoming more complicated. We serve warning that we shall use this remark later and treat the theorems it implies as if we had proved them.

Chapter II

TRANSFORMATIONS

§20. DEFINITIONS AND EXAMPLES OF LINEAR TRANSFORMATIONS

We come now to the objects that really make vector spaces interesting.

> DEFINITION. A <u>linear transformation</u> (or <u>operator</u>), A, on a vector space $\mathfrak{V}$ is a correspondence which assigns to every vector x in $\mathfrak{V}$ another vector, which we shall denote by Ax, in $\mathfrak{V}$, in such a way that, identically in the vectors x, y and scalars α, β, we have
>
> (1) $$A(\alpha x + \beta y) = \alpha Ax + \beta Ay.$$

We make again the remark that we made in connection with the definition of linear functionals, namely that for a linear transformation A, as we defined it, $A0 = 0$. For this reason such transformations are sometimes called <u>homogeneous</u> linear transformations.

Before discussing any properties of linear transformations we give several examples. We shall not bother to prove that the transformations we mention are indeed linear: in all cases the verification of the validity of (1) is a simple exercise.

(1) Two special transformations of considerable importance for the study that follows, and for which we shall consistently reserve the symbols 0 and 1 respectively, are defined, for every x, by $0x = 0$ and $1x = x$.

(2) Let x_0 be any fixed vector in $\mathfrak{V}$, and let $y_0 = y_0(x)$ be any linear functional on $\mathfrak{V}$; we define $Ax = y_0(x)x_0$. More generally: let $x_1, \ldots, x_n$ be an arbitrary finite set of vectors in $\mathfrak{V}$ and let $y_1, \ldots, y_n$ be n linear functionals on $\mathfrak{V}$; we define $Ax = y_1(x)x_1 + \cdots + y_n(x)x_n$. It is not difficult to prove that if, in particular, $\mathfrak{V}$ is n-dimensional, and $x_1, \ldots, x_n$ is a basis in $\mathfrak{V}$, then every linear transformation A has the form just described.

(3) Let $(j_1, \ldots, j_n)$ be any permutation of the first n positive integers; for any $x = \{\xi_1, \ldots, \xi_n\}$ in $\mathfrak{C}_n$, define $Ax = \{\xi_{j_1}, \ldots, \xi_{j_n}\}$. Similarly, let $j(t)$ be any polynomial, with real coefficients, in the real variable t; for every $x = x(t)$ in $\mathfrak{P}$, define $Ax = x(j(t))$.

(4) For any $x = x(t) = \sum_{j=0}^{n-1} \xi_j t^j$ in $\mathfrak{P}_n$, define $Dx = \sum_{j=0}^{n-1} j\,\xi_j t^{j-1}$. (We use the letter D here to remind the reader that Dx is the derivative of the polynomial $x = x(t)$. We remark that we may have defined D on $\mathfrak{P}$ as well as on $\mathfrak{P}_n$: we shall make use of this fact later. Observe that for polynomials the definition of derivation can be given purely algebraically, and does not need the usual theory of limiting processes.)

(5) For every $x = x(t) = \sum_{j=0}^{n-1} \xi_j t^j$ in $\mathfrak{P}$ (without subscript) we define $Jx = \sum_{j=0}^{n-1} \frac{\xi_j}{j+1} t^{j+1}$. (Once more we are disguising by algebraic notation a well known analytic concept. Just as in (4) Dx stood for dx/dt, so here Jx is the same as $\int_0^t x(s)ds$.)

(6) Let $m(t)$ be any polynomial, with complex coefficients, in the real variable t. (We may, although it is not particularly profitable to do so, consider $m(t)$ as an element of $\mathfrak{P}$). For every $x = x(t)$ in $\mathfrak{P}$ we define $Mx = m(t)x(t)$. For later purposes we introduce a special notation: in case $m(t) = t$, we shall write T for the operator M; thus $Tx = tx(t)$.

It is often useful to consider linear transformations

(such, for example, as we mentioned in our definition of isomorphism) from one vector space to another; in this book we restrict ourselves to the more special situation.

§21. LINEAR TRANSFORMATIONS AS A VECTOR SPACE

We proceed now to derive certain elementary properties of, and relations between, linear transformations on a vector space. More particularly we shall indicate several ways of making new transformations out of old ones; we shall generally be satisfied with giving the definition of the new transformations and we shall omit the proof of linearity.

If A and B are any two linear transformations, we define their <u>sum</u>, $S = A+B$, by the equation $Sx = Ax + Bx$ (for every x). We observe that the commutativity and associativity of addition in $\mathfrak{V}$ imply immediately that the addition of linear transformations is commutative and associative. Much more than this is true. If we consider the sum of any A and 0 (as defined in example (1), §20) we see that $A + 0 = A$. If for any A, we denote by $-A$ the transformation defined by $(-A)x = -(Ax)$, we see that $A + (-A) = 0$, and that $-A$, as so defined, is the only linear transformation B with the property that $A + B = 0$. To sum up: the properties of a vector space, described in axiom (A) of §1, appear again in the set of all linear transformations on the space; the set of all linear transformations is an abelian group with respect to the operation of addition.

We continue in the same spirit: it will not, by now, surprise anybody if the axiom (B) of vector spaces is also satisfied by the set of all linear transformations. It is. For any A, and any scalar α, we define the product of A by α, αA, by the equation $(\alpha A)x = \alpha(Ax)$. Axiom (B) is immediately verified; we sum up in the following theorem.

THEOREM. The set of all linear transformations on a vector space is itself a vector space.

We shall usually ignore this theorem; the reason is that we can say much more about linear transformations and the mere fact that they form a vector space is used only very rarely. The "much more" that we can say is that there exists for linear transformations a more or less decent definition of multiplication, which we discuss in the next section.

§22. PRODUCTS OF LINEAR TRANSFORMATIONS

The product P of two linear transformations A and B, $P = AB$, is defined by the equation $Px = A(Bx)$.

The notion of multiplication is fundamental for all that follows. Before giving any examples to illustrate the meaning of operator products, let us observe the implications of the symbolism, $P = AB$. P is a transformation; i.e. given a vector x, P does something to it. What it does is found out by operating on x with B, i.e. finding Bx, and then operating on the result with A. In other words if we look upon the symbol for an operator as a recipe for performing a certain act, then the product of two operators is to be read from right to left. $P = AB$ means: "Operate first with B, and then with A". This may seem like an undue amount of fuss to raise about a small point; however, as we shall soon see, operator multiplication is not, in general, commutative, $AB \neq BA$, and it makes a lot of difference in which order we operate.

The most notorious example of non commutative operators is found in the space $\mathfrak{P}$. We consider $Dx = dx/dt$ and $Tx = tx(t)$, (see §20); we have:

$$DTx = D(tx(t)) = x(t) + t(dx/dt),$$
$$TDx = T(dx/dt) = t(dx/dt).$$

In other words not only is it false that $DT = TD$ (so that $DT - TD = 0$), but in fact, for every x, $(DT - TD)x = x$, so that $DT - TD = 1$.

On the basis of the examples in §20 the reader should be able to construct examples of non commutative operators. Those who are used to thinking of linear transformations geometrically can, for example, readily convince themselves that the product of two rotations of $\mathcal{R}_3$ (about the origin) depends in general on the order in which they are performed.

The formal algebraic properties of numerical multiplication are most of them (with the already mentioned notable exception of commutativity) valid in the operator calculus. Thus we have:

(1) $A0 = 0A = 0$

(2) $A1 = 1A = A$

(3) $A(B + C) = AB + AC$

(4) $(A + B)C = AC + BC$

(5) $A(BC) = (AB)C.$

The proofs of all these identities are immediate consequences of the definitions of addition and multiplication; to illustrate the principle we prove (5), the associativity of multiplication. Let x be any vector, and denote by L and R the left and right sides of (5) respectively; we must show that $Lx = Rx$. We write $y = Cx$, $z = By$, $u = Az$; then $(BC)x = By = z$, so that

$$Lx = A(BC)x = Az = u,$$

$$Rx = (AB)Cx = ABy = Az = u.$$

§23. POLYNOMIALS IN A LINEAR TRANSFORMATION

The associative law of multiplication enables us to write the product of three (or more) factors without any brackets; in particular we may consider the product of any finite number, say m, of factors all equal to A,

$$\underbrace{AA \cdots A}_{m \text{ times}} .$$

This product depends only on A and m (and not, as we just remarked, on any bracketing of the factors); we shall denote it by A^m. The justification for this notation is that, although in general operator multiplication is not commutative, for the powers of one operator we do have the usual law of exponents, $A^nA^m = A^{n+m}$. We observe that $A^1 = A$; it is customary also to define $A^0 = 1$. With these definitions the calculus of powers of a single transformation is almost exactly the same as in ordinary arithmetic. We may in particular define polynomials in a linear transformation. Thus if $p(\tau) = \alpha_0 + \alpha_1\tau + \cdots + \alpha_n\tau^n$ is any polynomial in the variable τ (with scalar coefficients) we may form the linear transformation

$$p(A) = \alpha_0 1 + \alpha_1 A + \cdots + \alpha_n A^n.$$

The rules for algebraic manipulation of such polynomials are easy. Thus $p(\tau)q(\tau) = r(\tau)$ implies $p(A)q(A) = r(A)$, (so that, in particular, any $p(A)$ and $q(A)$ are commutative); if $p(\tau) \equiv 1$, then $p(A) = 1$; if $p(\tau) \equiv 0$, $p(A) = 0$; if $p(\tau) + q(\tau) = r(\tau)$ then $p(A) + q(A) = r(A)$.

It is not possible to give any sensible interpretation to $p(A,B)$, if $p(\sigma,\tau)$ is any polynomial in two variables, and A and B are any two linear transformations. The trouble, of course, is that A and B may not commute, and even a simple monomial, such as $\sigma^2\tau$, will cause confusion. If $p(\sigma,\tau) = \sigma^2\tau$, what should we mean by $p(A,B)$? Should it be A^2B, or ABA, or BA^2? It is important to realize that there is a difficulty here; fortunately for us it is not necessary to try to get around it. We shall work with polynomials in several variables only in connection with commutative transformations, and here everything is simple. We observe that if $AB = BA$ then $A^nB^m = B^mA^n$, and, therefore, $p(A,B)$ has a well defined meaning for every $p(\sigma,\tau)$. The formal properties of the correspondence between transformations and polynomials are just as valid for several variables as for one; we omit the details.

For an example of the possible behaviour of powers of a transformation we look at the differentiation operator D, on $\mathfrak{P}$ (or, just as well, in $\mathfrak{P}_n$, for some n) It is easy to see that for any positive integer k, and any $x = x(t)$ in $\mathfrak{P}$, $D^k x = d^k x/dt^k$. We observe, that whatever else D does, it lowers the degree of the polynomial on which it operates by exactly one unit (assuming of course that this degree is $\geqq 1$). Let $x = x(t)$ be a polynomial of degree $n-1$, say; what is $D^{n-1}x$? Or put it another way: what is the product of the two (commutative) operators D^k and D^{n-k+1}, (where k is any integer between 0 and $n-1$), considered on the space $\mathfrak{P}_n$? We mention this example to bring out the disconcerting fact implied by the answer to the last question: the product of two operators, neither of which is zero, may vanish. (Non zero operators whose product with some other non zero operator is zero are called <u>divisors of zero</u>.)

§24. INVERSE OF A LINEAR TRANSFORMATION

In each of the two preceding sections we gave an example: these two examples bring out the two nasty properties that the multiplication of linear transformations has, namely non commutativity and the existence of divisors of zero. We turn now to the more pleasant properties that linear transformations sometimes have.

It may happen that the linear transformation A has one or both of the following two very special properties.

(1) $x_1 \neq x_2$ implies $Ax_1 \neq Ax_2$.

(2) To every vector y there corresponds (at least) one vector x such that $Ax = y$.

If ever A has both these properties we shall say that A <u>has an inverse</u>, or that A is <u>non singular</u>, and we define a linear transformation, called the <u>inverse</u> of A and denoted by A^{-1}, as follows. If y is any vector we may (by (2)) find an x for which $Ax = y$. This x is, moreover, uniquely determined, since $x' \neq x$ implies (by (1)) that $Ax' \neq y = Ax$. We define $A^{-1}y$ to be x. To prove that A^{-1} is linear, we evaluate $A^{-1}(\alpha_1 y_1 + \alpha_2 y_2)$. If $Ax_1 = y_1$ and $Ax_2 = y_2$, then the linearity of A tells us that $A(\alpha_1 x_1 + \alpha_2 x_2) = \alpha_1 y_1 + \alpha_2 y_2$, so that $A^{-1}(\alpha_1 y_1 + \alpha_2 y_2) = \alpha_1 x_1 + \alpha_2 x_2 = \alpha_1 A^{-1} y_1 + \alpha_2 A^{-1} y_2$.

As a trivial example of a transformation which has an inverse we mention the identity transformation 1; clearly $1^{-1} = 1$. The transformation 0 does not have an inverse; it violates both the conditions (1) and (2) about as strongly as they can be violated.

It is immediate from the definition that for any A which has an inverse we have

$$AA^{-1} = A^{-1}A = 1;$$

we shall now show that these equations serve to characterize A^{-1}. More precisely:

THEOREM 1. If A, B, and C are linear transformations such that

$$AB = CA = 1,$$

then A has an inverse and $A^{-1} = B = C$.

PROOF. If $Ax_1 = Ax_2$, then $CAx_1 = CAx_2$, so that (since $CA = 1$) $x_1 = x_2$; in other words the first condition of the definition of an inverse is satisfied. The second condition is also satisfied, for if y is any vector and $x = By$, then $y = ABy = Ax$. Multiplying $AB = 1$ on the left, and $CA = 1$ on the right, by A^{-1}, we see that $A^{-1} = B = C$.

To show that neither $AB = 1$ nor $CA = 1$ is by itself sufficient to ensure the existence of A^{-1}, we

call attention to the differentiation and integration
operators D and J, defined in §20, (4) and (5). Although DJ = 1, neither D nor J has an inverse; D
violates (1), and J violates (2). Query: Which of
the other transformations defined in §20 have inverses?

In finite dimensional spaces the situation is much
simpler.

THEOREM 2. A linear transformation A
on a finite dimensional vector space $\mathfrak{V}$ has
an inverse if and only if Ax = 0 implies
x = 0, or, alternatively, if and only if every
y in $\mathfrak{V}$ can be written in the form y = Ax.

PROOF. If A^{-1} exists the condition is satisfied
-- this much is trivial. Suppose now that Ax = 0 implies x = 0. Then x' $\neq$ x'', i.e. x' - x'' $\neq$ 0, implies
A(x' - x'') $\neq$ 0, i.e. Ax' $\neq$ Ax''; this proves (1). To
prove (2), let $(x_1, x_2, \ldots, x_n)$ be a basis in $\mathfrak{V}$; we
assert that $(Ax_1, \ldots, Ax_n)$ is also a basis. According to Theorem 2, §7, we need only prove linear independ-
ence. But, $\sum_i \alpha_i Ax_i = 0$ means $A(\sum_i \alpha_i x_i) = 0$, and,
by hypothesis, this implies $\sum_i \alpha_i x_i = 0$; the linear
independence of the x_i now tells us that $\alpha_1 = \cdots =$
$\alpha_n = 0$. It follows, of course, that every vector y
may be written in the form $y = \sum_i \alpha_i Ax_i = A(\sum_i \alpha_i x_i)$.
Let us assume next that every y is an Ax, and
let $(y_1, \ldots, y_n)$ be any basis in $\mathfrak{V}$. Corresponding
to each y_i we may find a (not necessarily unique) x_i
for which $y_i = Ax_i$; we assert that $(x_1, \ldots, x_n)$ is
also a basis. For $\sum_i \alpha_i x_i = 0$ implies $\sum_i \alpha_i Ax_i =$
$\sum_i \alpha_i y_i = 0$, so that $\alpha_1 = \cdots = \alpha_n = 0$. Consequently
every x may be written in the form $x = \sum_i \alpha_i x_i$, and
Ax = 0 implies, as in the argument just given, that
x = 0.

THEOREM 3. If A^{-1} and B^{-1} exist then $(AB)^{-1}$ exists and $(AB)^{-1} = B^{-1}A^{-1}$; if A^{-1} exists and $\alpha \neq 0$ then $(\alpha A)^{-1}$ exists and $(\alpha A)^{-1} = (1/\alpha)A^{-1}$; if A^{-1} exists then so does $(A^{-1})^{-1}$ and $(A^{-1})^{-1} = A$.

PROOF. According to Theorem 1, it is sufficient to prove (for the first statement) that the product of AB with $B^{-1}A^{-1}$, in both orders, is the identity; this verification we leave to the reader. The proofs of both the remaining statements are identical in principle with this one; the last statement, for example, follows from the fact that the equations $AA^{-1} = A^{-1}A = 1$ are completely symmetric in A and A^{-1}.

We conclude our discussion of inverses with the following comment. In the spirit of the preceding section we may if we like define rational functions of A, whenever possible, by using A^{-1}. We shall not find it useful to do this except in one case: if A^{-1} exists, then we know that $(A^n)^{-1}$ also exists, for $n = 1, 2, \ldots$; we shall write A^{-n} for this latter transformation, so that $A^{-n} = (A^n)^{-1} = (A^{-1})^n$.

§25. DEFINITION OF MATRICES

Let us now pick up the loose threads: we have introduced the new concept of linear transformation and we must find out what it has to do with the old concepts of bases, linear functionals, etc.

One of the most important concepts in the study of operators on finite dimensional vector spaces is that of a matrix. Since this concept usually has no decent analog in infinite dimensional spaces, and since it is possible in most considerations to do without it, we shall try not to use it in proving theorems. It is, however, important to know what a matrix is: we enter now into the detailed discussion.

DEFINITION. Let $\mathfrak{V}$ be an n dimensional vector space and let $\mathfrak{X} = (x_1, \ldots, x_n)$ be any basis in $\mathfrak{V}$; let A be a linear transformation on $\mathfrak{V}$. Since every vector is a linear combination of the x_i, we have in particular

(1) $$Ax_j = \sum_i \alpha_{ij} x_i$$

for $j = 1, \ldots, n$. The set (α_{ij}) of n^2 scalars, indexed with the double subscript i, j, is the <u>matrix of A in the coordinate system $\mathfrak{X}$</u>; we shall generally denote it by $[A]$, or, if it becomes necessary to indicate the particular basis $\mathfrak{X}$ under consideration, by $[A; \mathfrak{X}]$. A matrix (α_{ij}) is usually written in the form of a square array:

$$[A] = \begin{bmatrix} \alpha_{11} & \alpha_{12} & \cdots & \alpha_{1n} \\ \alpha_{21} & \alpha_{22} & \cdots & \alpha_{2n} \\ \cdot\cdot & \cdot\cdot & \cdots & \cdot\cdot \\ \alpha_{n1} & \alpha_{n2} & & \alpha_{nn} \end{bmatrix};$$

the scalars $(\alpha_{i1}, \ldots, \alpha_{in})$ form a <u>row</u>, and $(\alpha_{1j}, \ldots, \alpha_{nj})$ a <u>column</u>, of $[A]$.

The above definition does not define "matrix"; it defines "the matrix associated under certain conditions with a linear transformation". It is often useful to consider a matrix as something existing in its own right as a square array of scalars; in general, however, a matrix in this book will be tied up with a linear transformation and a basis. Rectangular (not square) matrices are usually considered in connection with linear transformations from one vector space to another; since we do not treat such transformations we shall also not discuss such matrices. Their theory is not very different from the theory we shall develop; in particular the reader will find it useful, after he has read the next section treating the square case, to try to define the algebraic operations (sum, product, etc.) for rectangular matrices.

We comment on notation. It is customary to use the same symbol, say A, for the matrix as for the transformation. The justification for this is to be found in the discussion below (of properties of matrices). We do not follow this custom here because one of our principal aims, in connection with matrices, is to emphasize that they depend on the coordinate system, (whereas the notion of linear transformation does not), and to study how the relation between matrices and operators changes as we pass from one coordinate system to another.

We call attention also to a peculiarity of the indexing of the elements α_{ij} of a matrix [A]. A basis is a basis, and so far, although we usually indexed its elements with the first n positive integers, the order of elements in it was entirely immaterial. It is customary, however, when speaking of matrices, to refer to, say, the first row or first column. This language is justified only if we think of the elements of the basis $\mathfrak{X}$ as arranged in a definite order. Since in the majority of our considerations the order of the rows and columns of a matrix is as irrelevant as the order of the elements of a basis, we did not include this aspect of matrices in our definition. It is important to realize that the appearance of the square array associated with [A] varies with the ordering of $\mathfrak{X}$. Everything we shall say about matrices (with very few exceptions, occuring mostly in the theory of direct products) can, accordingly, be interpreted from two different points of view; either in strict accordance with the letter of our definition, or else following a modified definition which makes correspond a matrix (with ordered rows and columns) not merely to a linear transformation and a basis, but

The relation between operators and matrices is exactly the same as the relation between vectors and their coordinates, and the analog of the isomorphism theorem of §8 is true in the best possible sense. We shall make these statements precise.

With the aid of a fixed basis $\mathfrak{X}$ we have made correspond a matrix [A] to every linear transformation A: the correspondence is described by the relations $Ax_j = \sum_i \alpha_{ij} x_i$, (§25, (1)). We assert now that this correspondence is one to one, in the sense that the matrices of two different operators are different, and every array (α_{ij}), of n^2 scalars is the matrix of some operator. To prove this, we observe in the first place that knowledge of the matrix of A completely determines A, (i.e. that Ax is thereby uniquely defined for every x), according to the relations:

(5) $$x = \sum_j \xi_j x_j$$

(6) $$Ax = \sum_j \xi_j Ax_j = \sum_j \xi_j (\sum_i \alpha_{ij} x_i) = \sum_i (\sum_j \alpha_{ij} \xi_j) x_i.$$

(In other words if $y = Ax = \sum_i \eta_i x_i$, then

(7) $$\eta_i = \sum_j \alpha_{ij} \xi_j.$$

Compare this with §25, (2), and the subsequent comments on the perversity of indices). In the second place, there is no law against reading the relation $Ax_j = \sum_i \alpha_{ij} x_i$ backwards: if, in other words, (α_{ij}) is any array, we may write this relation as the definition of a linear transformation A; it is clear that the matrix of A will be exactly (α_{ij}). (We emphasize, however, once more the fundamental fact that this one to one correspondence between operators and matrices was set up by means of a particular coordinate system, and that, as we pass from one coordinate system to another, the same linear transformation may correspond to several matrices, and one matrix may be the correspondent of many linear transformations).

We sum up.

THEOREM. Among the set of all matrices (α_{ij}), (β_{ij}), etc., $i,j = 1, \ldots, n$, (not considered in relation to linear transformations), we define sum, scalar multiplication, product, (o_{ij}), and (e_{ij}), by

$$(\alpha_{ij}) + (\beta_{ij}) = (\alpha_{ij} + \beta_{ij}),$$
$$\alpha(\alpha_{ij}) = (\alpha\alpha_{ij})$$
$$(\alpha_{ij})(\beta_{ij}) = (\textstyle\sum_k \alpha_{ik}\beta_{kj}).$$
$$o_{ij} = 0, \quad e_{ij} = \delta_{ij}.$$

Then the correspondence, (established by means of an arbitrary coordinate system $\mathfrak{X} = (x_1, \ldots, x_n)$ of the n-dimensional vector space $\mathfrak{V}$), between all linear transformations A on $\mathfrak{V}$ and all matrices (α_{ij}), described by $Ax_j = \sum_i \alpha_{ij}x_i$, is an isomorphism: in other words it is a one to one correspondence that preserves sum, scalar multiplication, product, 0, and 1.

We have carefully avoided discussing the matrix of A^{-1}. It is possible to give an expression for $[A^{-1}]$ in terms of the elements α_{ij} of $[A]$, but the expression is not simple and, for us, fortunately, not useful.

§27. REDUCIBILITY

A possible relation between linear manifolds $\mathfrak{M}$ in a vector space $\mathfrak{V}$ and linear transformations A on $\mathfrak{V}$ is that of invariance. $\mathfrak{M}$ is <u>invariant</u> under A if x in $\mathfrak{M}$ implies that Ax is in $\mathfrak{M}$. (Observe that the implication relation is required in only one direction: we do not assume that every y in $\mathfrak{M}$ can be written in the form $y = Ax$ with x in $\mathfrak{M}$; we do not even assume that Ax in $\mathfrak{M}$ implies x in $\mathfrak{M}$. See example below). Another locution for the same concept is: $\mathfrak{M}$ <u>reduces</u> A. (Reducibility is often defined for sets of linear transformations as well as for a single one: $\mathfrak{M}$ reduces a set if it reduces each member of the set). We know that a linear manifold $\mathfrak{M}$ in a vector space is itself a vector space: if we know that $\mathfrak{M}$ reduces A we may ignore the fact that A is defined outside $\mathfrak{M}$ and we may consider A as a linear transformation defined on the vector space $\mathfrak{M}$.

What can be said about the matrix of an operator A, on an n-dimensional vector space $\mathfrak{V}$, which is reduced by some $\mathfrak{M}$? In other words: is there a clever way of selecting a basis $\mathfrak{X} = (x_1, \ldots, x_n)$ in $\mathfrak{V}$ so that $[A] = [A; \mathfrak{X}]$ will have some particularly simple form? The answer is in Theorem 2 of §11: we may choose $\mathfrak{X}$ so that $x_1, \ldots, x_m$ are in $\mathfrak{M}$ and $x_{m+1}, \ldots, x_n$ are not. Let us express Ax_j in terms of $x_1, \ldots, x_n$. For $m+1 \leqq j \leqq n$ there is not much we can say: $Ax_j = \sum_i \alpha_{ij}x_i$. For $1 \leqq j \leqq m$, however, x_j is in $\mathfrak{M}$, and therefore (since $\mathfrak{M}$ is invariant under A) Ax_j is in $\mathfrak{M}$. Consequently Ax_j is in this case a linear combination of $x_1, \ldots, x_m$; the α_{ij} with $m+1 \leqq i \leqq n$ are zero. Hence the matrix [A] of A, in this coordinate system, will have the form

$$(1) \quad [A] = \begin{bmatrix} [A_1] & [B_0] \\ [0] & [A_2] \end{bmatrix},$$

where $[A_1]$ is the (m rowed) matrix of A when considered as a linear transformation on the space $\mathfrak{M}$, with the coordinate system $x_1, \ldots, x_m$, $[A_2]$ and $[B_0]$ are some arrays of numbers (in size (n-m) times (n-m) and m times (n-m) respectively), and [0] denotes the rectangular ((n-m) times m) array consisting of zeros only. (It is important to observe the unpleasant fact that

$[B_0]$ need not be zero).

For an example we may consider the differentiation operator D on the space $\mathfrak{P}_n$, and the linear manifold $\mathfrak{M}$ spanned by the vectors $1, t, t^2, \ldots, t^m$, $1 \leq m < n$. We leave it to the reader to verify that in this case $\mathfrak{M}$ is indeed invariant under D, but all the unpleasant possibilities we have been hinting at ($[B_0] \neq [0]$, D singular in $\mathfrak{M}$, etc.) become actualities.

§28. COMPLETE REDUCIBILITY AND DIRECT SUMS OF TRANSFORMATIONS

A particularly important subcase of the notion of reducibility is that of complete reducibility. If $\mathfrak{M}$ and $\mathfrak{N}$ are two linear manifolds such that both are invariant under A and such that $\mathfrak{V}$ is their direct sum, then A is <u>completely reduced</u> (decomposed) by the pair $(\mathfrak{M}, \mathfrak{N})$. (The difference between complete reducibility and just plain reducibility is that in the latter case among the collection of all linear manifolds invariant under A we may not be able to pick out any two, other than $\mathfrak{O}$ and $\mathfrak{V}$, with the property that $\mathfrak{V}$ is their direct sum. Or, saying it the other way, if $\mathfrak{M}$ is invariant under A, there are, to be sure, many ways of finding an $\mathfrak{N}$ such that $\mathfrak{V} = \mathfrak{M} \oplus \mathfrak{N}$, but it may happen that no such $\mathfrak{N}$ will be invariant under A).

The process described above may also be turned around. Let $\mathfrak{M}$ and $\mathfrak{N}$ be any two vector spaces, and let A and B be any two linear transformations (on $\mathfrak{M}$ and $\mathfrak{N}$ respectively). Let $\mathfrak{V}$ be the direct sum $\mathfrak{M} \oplus \mathfrak{N}$; we may define on $\mathfrak{V}$ a linear transformation C, called the direct sum of A and B, defined by

$$Cz = C\{x,y\} = \{Ax, By\}.$$

We shall omit the detailed discussion of direct sums of transformations: we shall merely mention the results. Their proofs are easy. If $(\mathfrak{M}, \mathfrak{N})$ completely reduces

C, and if we denote by A the linear transformation C considered on $\mathfrak{M}$ alone, and by B the linear transformation C considered on $\mathfrak{N}$ alone, then C is the direct sum of A and B. By suitable choice of basis -- namely by choosing $x_1, \ldots, x_m$ in $\mathfrak{M}$ and $x_{m+1}, \ldots, x_n$ in $\mathfrak{N}$ -- the matrix of the direct sum of A and B will have the form (1) of §27, with $[A_1] = [A]$, $[B_0] = [0]$, and $[A_2] = [B]$. If $p(\tau)$ is any polynomial, and if we write $A' = p(A)$, $B' = p(B)$, then the direct sum, C', of A' and B', will be $p(C)$.

§29. PROJECTIONS

More important for our purposes is another connection between direct sums and linear transformations.

DEFINITION. If $\mathfrak{V}$ is the direct sum of $\mathfrak{M}$ and $\mathfrak{N}$, so that every z in $\mathfrak{V}$ may be written, uniquely, in the form $z = x+y$, with x in $\mathfrak{M}$ and y in $\mathfrak{N}$, we define a transformation E by $Ez = x$. E is called the <u>projection</u> on $\mathfrak{M}$ along $\mathfrak{N}$. $V = M \oplus N$

Granting that direct sums are important, projections are also, since, as we shall see, they are a very powerful algebraic tool in studying the geometric concept of direct sum. (The reader will easily satisfy himself about the reason for the word "projection", by drawing a pair of axes (linear manifolds) in the plane (their direct sum). To make the picture look general enough do not draw perpendicular axes!)

We skipped over one point whose proof is easy enough to skip over, but whose existence should be recognized: it must be shown that E is a <u>linear</u> transformation. We leave this verification to the reader, and go on to look for special properties of projections.

THEOREM 1. A linear transformation E is a projection on some linear manifold $\mathfrak{M}$ if and only if it is idempotent, i.e. $E^2 = E$.

PROOF. If E is the projection on $\mathfrak{M}$ along $\mathfrak{N}$, and if $z = x+y$ is the decomposition of z, with x in $\mathfrak{M}$ and y in $\mathfrak{N}$, then the decomposition of x is $x+0$, so that

$$E^2z = EEz = Ex = x = Ez.$$

Conversely, suppose that $E^2 = E$. Let $\mathfrak{N}$ be the set of all vectors z in $\mathfrak{V}$ for which $Ez = 0$; let $\mathfrak{M}$ be the set of all vectors z for which $Ez = z$. It is clear that both $\mathfrak{M}$ and $\mathfrak{N}$ are linear manifolds: we shall prove that $\mathfrak{V} = \mathfrak{M} \oplus \mathfrak{N}$. In view of the theorem of §17 we need to prove that $\mathfrak{M}$ and $\mathfrak{N}$ are disjoint and that together they span $\mathfrak{V}$.

If z is in $\mathfrak{M}$, $Ez = z$; if z is in $\mathfrak{N}$, $Ez = 0$; hence if z is in both $\mathfrak{M}$ and $\mathfrak{N}$, $z = 0$. For an arbitrary z we have

$$z = Ez + (1-E)z.$$

If we write $Ez = x$ and $(1-E)z = y$, then $Ex = E^2z = Ez = x$, and $Ey = E(1-E)z = Ez-E^2z = 0$, so that x is in $\mathfrak{M}$ and y is in $\mathfrak{N}$. This proves that $\mathfrak{V} = \mathfrak{M} \oplus \mathfrak{N}$, and that the projection on $\mathfrak{M}$ along $\mathfrak{N}$ is precisely E.

As an immediate consequence of the above proof we obtain also:

THEOREM 2. If E is the projection on $\mathfrak{M}$ along $\mathfrak{N}$, then $\mathfrak{M}$ and $\mathfrak{N}$ are, respectively, the sets of all solutions of the equations $Ez = z$, and $Ez = 0$.

By means of these two theorems we can remove the apparent asymmetry in the definition of projections between the roles played by $\mathfrak{M}$ and $\mathfrak{N}$. If to every $z = x+y$

we make correspond not x, but y, we also get an idempotent linear transformation. This transformation (namely $1-E$) is the projection on $\mathfrak{N}$ along $\mathfrak{M}$. Hence:

THEOREM 3. E is a projection if and only if $1-E$ is a projection; if E is the projection on $\mathfrak{M}$ along $\mathfrak{N}$, then $1-E$ is the projection on $\mathfrak{N}$ along $\mathfrak{M}$.

Query: what is the necessary and sufficient condition on $y_0(x)$, in the first example of (2), §20, in order that the A defined there be a projection?

§30. ALGEBRAIC COMBINATIONS OF PROJECTIONS

Continuing in the spirit of Theorem 3 of the preceding section we investigate conditions under which various algebraic combinations of projections are themselves projections.

THEOREM. We assume that E_1 and E_2 are projections on $\mathfrak{M}_1$ and $\mathfrak{M}_2$ along $\mathfrak{N}_1$ and $\mathfrak{N}_2$ respectively. We make three assertions.

(i) $E_1 + E_2$ is a projection if and only if $E_1E_2 = E_2E_1 = 0$; if this condition is satisfied then $E = E_1 + E_2$ is the projection on $\mathfrak{M}$ along $\mathfrak{N}$, where $\mathfrak{M} = \mathfrak{M}_1 \oplus \mathfrak{M}_2$ and $\mathfrak{N} = \mathfrak{N}_1 \cap \mathfrak{N}_2$.

(ii) $E_1 - E_2$ is a projection if and only if $E_1E_2 = E_2E_1 = E_2$; if this condition is satisfied then $F = E_1 - E_2$ is the projection on $\mathfrak{M}$ along $\mathfrak{N}$, where $\mathfrak{M} = \mathfrak{M}_1 \cap \mathfrak{N}_2$, and $\mathfrak{N} = \mathfrak{N}_1 \oplus \mathfrak{M}_2$.

(iii) If $E_1E_2 = E_2E_1 = E$ then E is the projection on $\mathfrak{M}$ along $\mathfrak{N}$ where $\mathfrak{M} = \mathfrak{M}_1 \cap \mathfrak{M}_2$ and $\mathfrak{N} = \mathfrak{N}_1 + \mathfrak{N}_2$.

PROOF. We recall the notation used: for linear manifolds $\mathfrak{H}$ and $\mathfrak{K}$, $\mathfrak{H} + \mathfrak{K}$ denotes the linear manifold spanned by $\mathfrak{H}$ and $\mathfrak{K}$; writing $\mathfrak{H} \oplus \mathfrak{K}$ implies that $\mathfrak{H}$ and $\mathfrak{K}$ are disjoint, and then $\mathfrak{H} \oplus \mathfrak{K} = \mathfrak{H} + \mathfrak{K}$; and $\mathfrak{H} \cap \mathfrak{K}$ is the intersection of $\mathfrak{H}$ and $\mathfrak{K}$.

(i) If $E_1 + E_2 = E$ is a projection then $(E_1 + E_2)^2 = E^2 = E = E_1 + E_2$, whence the cross product terms must disappear:

$$E_1E_2 + E_2E_1 = 0. \tag{1}$$

If we multiply (1) on both right and left by E_1 we obtain

$$E_1E_2 + E_1E_2E_1 = 0, \tag{2}$$

$$E_1E_2E_1 + E_2E_1 = 0; \tag{3}$$

subtracting we get $E_1E_2 - E_2E_1 = 0$. Hence E_1 and E_2 are commutative, and (1) implies that their product is zero. Since, conversely, $E_1E_2 = E_2E_1 = 0$ clearly implies (1) we see that the condition is also sufficient to ensure that E be a projection.

Let us suppose, from now on, that E is a projection; by Theorem 2 of §29, $\mathfrak{M}$ and $\mathfrak{N}$ are, respectively, the sets of all solutions of the equations $Ez = z$ and $Ez = 0$. Let us write $z = x_1 + y_1 = x_2 + y_2$, where $x_i = E_iz$ is in $\mathfrak{M}_i$ and $y_i = (1-E_i)z$ is in $\mathfrak{N}_i$, $i = 1,2$. If z is in $\mathfrak{M}$, $E_1z + E_2z = z$, then

$$z = E_1(x_2+y_2) + E_2(x_1+y_1) = E_1y_2 + E_2y_1.$$

Since $E_1(E_1y_2) = E_1y_2$ and $E_2(E_2y_1) = E_2y_1$, we have exhibited z as a sum of a vector from $\mathfrak{M}_1$ and a vector from $\mathfrak{M}_2$, so that $\mathfrak{M} \subset \mathfrak{M}_1 + \mathfrak{M}_2$. Conversely, if z is a sum of a vector from $\mathfrak{M}_1$ and a vector from $\mathfrak{M}_2$, then $(E_1 + E_2)z = z$, so that z is in $\mathfrak{M}$, and consequently $\mathfrak{M} = \mathfrak{M}_1 + \mathfrak{M}_2$. Finally, if z belongs to both $\mathfrak{M}_1$ and $\mathfrak{M}_2$, $E_1z = E_2z = z$, then $z = E_1z = E_1(E_2z) = 0$, so that $\mathfrak{M}_1$ and $\mathfrak{M}_2$ are disjoint, and $\mathfrak{M} = \mathfrak{M}_1 \oplus \mathfrak{M}_2$.

It remains to find $\mathfrak{N}$, i.e. to find all solutions of $E_1 z + E_2 z = 0$. If z is in $\mathfrak{N}_1 \cap \mathfrak{N}_2$ this equation is clearly satisfied; conversely $E_1 z + E_2 z = 0$ implies (upon multiplication on the left by E_1 and E_2 respectively) that $E_1 z + E_1 E_2 z = 0$ and $E_2 E_1 z + E_2 z = 0$. Since $E_1 E_2 z = E_2 E_1 z = 0$ for all z, we obtain finally $E_1 z = E_2 z = 0$, so that z belongs to both $\mathfrak{N}_1$ and $\mathfrak{N}_2$.

Using the technique and the results obtained in this proof, the proofs of the remaining parts of the theorem are easy.

(ii) According to Theorem 3 of §29, $E_1 - E_2$ is a projection if and only if $1 - (E_1 - E_2) = (1 - E_1) + E_2$ is a projection. According to (i) this happens (since, of course, $1 - E_1$ is a projection on $\mathfrak{N}_1$ along $\mathfrak{M}_1$) if and only if

$$(4) \qquad (1 - E_1)E_2 = E_2(1 - E_1) = 0,$$

and in this case $(1 - E_1) + E_2$ is the projection on $\mathfrak{N}_1 \oplus \mathfrak{M}_2$ along $(\mathfrak{M}_1 \cap \mathfrak{N}_2)$. Since (4) is equivalent to $E_1 E_2 = E_2 E_1 = E_2$, the proof of (ii) is complete.

(iii) That $E = E_1 E_2 = E_2 E_1$ implies that E is a projection is clear, since E is idempotent. We assume, therefore, that E_1 and E_2 commute and we find $\mathfrak{M}$ and $\mathfrak{N}$. If $Ez = z$, then $E_1 z = E_1 Ez = E_1 E_1 E_2 z = E_1 E_2 z = z$, and similarly $E_2 z = z$, so that z is contained in both $\mathfrak{M}_1$ and $\mathfrak{M}_2$. The converse is clear: if $E_1 z = z = E_2 z$, then $Ez = z$. Suppose next that $E_1 E_2 z = 0$; it follows that $E_2 z$ belongs to $\mathfrak{N}_1$, and, from the commutativity of E_1 and E_2, that $E_1 z$ belongs to $\mathfrak{N}_2$. This is more symmetry than we need; since $z = E_2 z + (1 - E_2)z$, and since $(1 - E_2)z$ is in $\mathfrak{N}_2$, we have exhibited z as a sum of a vector from $\mathfrak{N}_1$ and a vector from $\mathfrak{N}_2$. Conversely if z is such a sum then $E_1 E_2 z = 0$; this concludes the proof that $\mathfrak{N} = \mathfrak{N}_1 + \mathfrak{N}_2$.

We shall return to theorems of this type later and we shall obtain, in certain cases, more precise results. Before leaving the subject, however, we call attention to a few minor peculiarities of the theorem of this section. We observe first that although in both (i) and (ii) one of $\mathfrak{M}$ and $\mathfrak{N}$ was a <u>direct</u> sum of two of the given linear manifolds, in (iii) we only stated $\mathfrak{N} = \mathfrak{N}_1 + \mathfrak{N}_2$. Consideration of the possibility $E_1 = E_2 = E$ shows that this is unavoidable. Also: the condition of (iii) was asserted to be sufficient only; it is possible to construct projections E_1 and E_2 whose product E_1E_2 is a projection, but for which E_1E_2 and E_2E_1 are different. Finally, it may be conjectured that it is possible to extend, by induction, the result of (i) to more than two summands. Although this is true it is surprisingly non trivial; we shall prove it later in a special case of interest.

§31. APPLICATION TO REDUCIBILITY AND INVOLUTIONS

We have already seen that the study of projections is equivalent to the study of direct sum decompositions. By means of projections we may also study the notions of reducibility and complete reducibility.

THEOREM 1. If a linear manifold $\mathfrak{M}$ is invariant under the linear transformation A (i.e. A is reduced by $\mathfrak{M}$) then EAE = AE for every projection E on $\mathfrak{M}$. Conversely, if EAE = AE is valid for some projection E on $\mathfrak{M}$ then $\mathfrak{M}$ reduces A.

PROOF. Suppose that A is reduced by $\mathfrak{M}$ and that $\mathfrak{V} = \mathfrak{M} \oplus \mathfrak{N}$ for some $\mathfrak{N}$; let E be the projection on $\mathfrak{M}$ along $\mathfrak{N}$. Then for any $z = x+y$ (with x in $\mathfrak{M}$, y in $\mathfrak{N}$) we have AEz = Ax, and EAEz = EAx;

this last term is, however, again equal to Ax since x being in $\mathfrak{M}$ guarantees the presence of Ax in $\mathfrak{M}$.

Conversely, suppose that $\mathfrak{V} = \mathfrak{M} \oplus \mathfrak{N}$, and that EAE = AE for the projection E on $\mathfrak{M}$ along $\mathfrak{N}$. If x is in $\mathfrak{M}$, then Ex = x, so that

$$EAx = EAEx = AEx = Ax,$$

and consequently Ax is also in $\mathfrak{M}$.

THEOREM 2. If $\mathfrak{M}$ and $\mathfrak{N}$ are two linear manifolds with $\mathfrak{V} = \mathfrak{M} \oplus \mathfrak{N}$, the linear transformation A is completely reduced by the pair ($\mathfrak{M}$, $\mathfrak{N}$) if and only if EA = AE, where E is the projection on $\mathfrak{M}$ along $\mathfrak{N}$.

PROOF. We first assume EA = AE and we prove that A is completely reduced by ($\mathfrak{M}$, $\mathfrak{N}$). If x is in $\mathfrak{M}$ then Ax = AEx = EAx, so that Ax is also in $\mathfrak{M}$; if x is in $\mathfrak{N}$ then Ex = 0, and EAx = AEx = A0 = 0 so that Ax is also in $\mathfrak{N}$.

Finally we assume that A is completely reduced by ($\mathfrak{M}$, $\mathfrak{N}$), and we prove that EA = AE. Since A is reduced by $\mathfrak{M}$, Theorem 1 assures us that EAE = AE; since A is also reduced by $\mathfrak{N}$, and since 1 - E is a projection on $\mathfrak{N}$, we have (1-E)A(1-E) = A(1-E), whence (after carrying out the indicated multiplications and simplifying) EAE = EA. This concludes the proof of the theorem. We observe that the first part of the theorem could also have been deduced from Theorem 1 by formal manipulation and it was only for the sake of variety that we chose our method. The interested reader might carry out the suggested proof.

We conclude, for the present, our discussion of projections with two isolated comments of some interest.

(A). There is an amusing connection between the

idempotent operators just studied (i.e. those satisfying the equation $E^2 = E$) and involutions (i.e. operators U satisfying the equation $U^2 = 1$). Let us write

(1) $$U = 2E - 1,$$

(2) $$E = \frac{1}{2} (U + 1).$$

We assert: the formulae (1) and (2) establish a one to one correspondence between all idempotents E and all involutions U; if U corresponds to E by (1) then E corresponds to U by (2). To prove this we must show that for any idempotent E the U defined by (1) is an involution, and, similarly, that for any involution U the E defined by (2) is idempotent. These verifications (which consist of squaring the right sides of (1) and (2) and substituting E and 1 for E^2 and U respectively) we leave to the reader.

(B) The matrices associated with projections in finite dimensional spaces have, after proper choice of coordinates, very simple forms. Let $\mathfrak{B}$ be an n-dimensional vector space, and let E be the projection on a linear manifold $\mathfrak{M}$ along $\mathfrak{N}$. We may choose a coordinate system $x_1, \ldots, x_n$ in $\mathfrak{B}$ in such a way that $x_1, \ldots, x_m$ are in $\mathfrak{M}$ and $x_{n+1}, \ldots, x_n$ are in $\mathfrak{N}$. In this coordinate system the matrix $[E] = (e_{ij})$ of E will have the property that $e_{ij} = 0$ if $i \neq j$, and $e_{ii} = 1$ or 0 according as $i \leq m$ or $i > m$. Query: what does this result imply about involutions ?

§32. ADJOINT OPERATORS

Let us study next the relation between the notions of linear transformation and conjugate space. Let $\mathfrak{B}$ be any vector space and let y be any element of $\mathfrak{B}'$; for any linear transformation A on $\mathfrak{B}$ we consider the expression $[Ax,y]$. For each fixed $y, y' = y'(x) = [Ax,y]$ is a linear functional defined on $\mathfrak{B}$; using the square bracket notation for y' as well as for y we have $[Ax,y] = [x,y']$. If now we allow y to vary over $\mathfrak{V}'$, then this procedure makes correspond to each y a y' depending, of course, on y; we write $y' = A'y$. The defining property of A' is

(1) $$[Ax,y] = [x,A'y].$$

We assert that A' is a linear transformation on $\mathfrak{V}'$; for if $y = \alpha_1 y_1 + \alpha_2 y_2$, then

$$\begin{aligned}[x,A'y] = [Ax,y] &= \alpha_1[Ax,y_1] + \alpha_2[Ax,y_2] \\ &= \alpha_1[x,A'y_1] + \alpha_2[x,A'y_2] \\ &= [x, \alpha_1 A'y_1 + \alpha_2 A'y_2].\end{aligned}$$

The linear transformation A' is called the adjoint of A; we dedicate this section and the next to studying properties of A'. Let us first get the formal algebraic rules out of the way; they are the following.

(2) $$0' = 0$$

(3) $$1' = 1$$

(4) $$(A + B)' = A' + B'$$

(5) $$(\alpha A)' = \alpha A'$$

(6) $$(AB)' = B'A'$$

(7) $$(A^{-1})' = (A')^{-1}$$

Here (7) is to be interpreted in the following sense: if A has an inverse then A' also has an inverse and (7) is valid. The proofs of all these relations are elementary: to indicate the procedure we carry out the computations for (6) and (7). (6) is proved by the relations

$$[ABx,y] = [Bx,A'y] = [x,B'A'y].$$

To prove (7), suppose that A has an inverse; then $AA^{-1} = A^{-1}A = 1$. Applying to this relation (3) and (6) we obtain

$$(A^{-1})'A' = A'(A^{-1})' = 1;$$

Theorem 1 of §24 implies that $(A')^{-1}$ exists and is equal to $(A^{-1})'$.

In finite dimensional spaces (or, more properly speaking, in reflexive spaces) another important relation holds:

(8) $$A'' = A.$$

This relation has to be read with a grain of salt. As it stands A'' is an operator not on $\mathfrak{B}$ but on the conjugate space $\mathfrak{B}''$ of $\mathfrak{B}'$. If however, we identify $\mathfrak{B}''$ and $\mathfrak{B}$ according to the natural isomorphism, then A'' operates on $\mathfrak{B}$ and (8) makes sense. In this interpretation the proof of (8) is trivial. Since $\mathfrak{B}$ is reflexive we obtain every linear functional on $\mathfrak{B}'$ by considering $[x,y]$ as a function of y, with x fixed in $\mathfrak{B}$. Then $[x,A'y]$ is also a function (a linear functional) of y, and may therefore be written in the form $[x',y]$. The vector x' is, by definition, $A''x$. Hence we have, for every y in $\mathfrak{B}'$ and x in $\mathfrak{B}$

(9) $$[Ax,y] = [x,A'y] = [A''x,y]:$$

the equality of the first and last terms of (9) proves (8).

Under the hypothesis of (8) (i.e. reflexivity) the asymmetry in the interpretation of (7) may be removed: we assert that in this case the existence of $(A')^{-1}$ implies that of A^{-1} and, therefore, the validity of (7). Proof: we may apply the old interpretation of (7) to A' and A'' in place of A and A'.

Our discussion is summed up, in the reflexive finite dimensional case, by the assertion that $A \longrightarrow A'$ is a one to one algebraically anti-isomorphic mapping of the set of all linear operators on $\mathfrak{B}$ onto the set of all linear operators on $\mathfrak{B}'$. (The prefix "anti-" got attached because of the commutation rule (6)).

§33. ADJOINT OF A PROJECTION

There is one important case in which multiplication does not get turned around, i.e. when $(AB)' = A'B'$; namely, the case when A and B commute. In particular we have $(A^n)' = (A')^n$ and, more generally, for any polynomial $p(\tau)$, $(p(A))' = p(A')$. It follows from this that if E is a projection then so is E'. The question arises: what direct sum decomposition is E' associated with.

THEOREM 1. If E is the projection on $\mathfrak{M}$ along $\mathfrak{N}$, then E' is the projection on $\mathfrak{N}^0$ along $\mathfrak{M}^0$.

PROOF. That $(E')^2 = E'$ and that $\mathfrak{V}' = \mathfrak{N}^0 \oplus \mathfrak{M}^0$, we have already seen. (Cf. §19). It is necessary only to find the linear manifolds of solutions of $E'y = 0$ and $E'y = y$. This we do in four steps.

(i) If y is in $\mathfrak{M}^0$ then for all x

$$[x,E'y] = [Ex,y] = 0$$

so that $E'y = 0$.

(ii) If $E'y = 0$ then for all x in $\mathfrak{M}$

$$[x,y] = [Ex,y] = [x,E'y] = 0$$

so that y is in $\mathfrak{M}^0$.

(iii) If y is in $\mathfrak{N}^0$ then for all x

$$[x,y] = [Ex,y] + [(1-E)x,y] = [Ex,y] = [x,E'y]$$

so that $E'y = y$.

(iv) If $E'y = y$ then for all x in $\mathfrak{N}$

$$[x,y] = [x,E'y] = [Ex,y] = 0$$

so that y is in $\mathfrak{N}^0$.

Steps (i) and (ii) together show that the set of solutions of $E'y = 0$ is precisely $\mathfrak{M}^0$; steps (iii) and (iv) together show that the set of solutions of $E'y = y$ is precisely $\mathfrak{N}^0$. This concludes the proof of the theorem.

THEOREM 2. If $\mathfrak{M}$ reduces A then $\mathfrak{M}^0$ reduces A'; if A is completely reduced by

($\mathfrak{M}$, $\mathfrak{N}$) then A' is completely reduced
by ($\mathfrak{M}^{\circ}$, $\mathfrak{N}^{\circ}$)

PROOF. We shall prove only the first statement; the
second one clearly follows from it. We first observe the
following identity, valid for any three operators E, F,
and A, subject to the relation $F = 1 - E$:
(1) $FAF - FA = EAE - AE$.
(Compare this with the proof of Theorem 2, §31). Let E
be any projection on $\mathfrak{M}$; by Theorem 1, §31, the right
member of (1) vanishes, and so, therefore, does the left
member. By taking adjoints we obtain $F'A'F' = A'F'$;
since by Theorem 1 of the present section $F' = E'$ is a
projection on $\mathfrak{M}^{\circ}$, the proof of Theorem 2 is complete.

We conclude by discussing the matrices of adjoint
operators; this discussion is meant to illuminate the
entire theory and to enable the reader to construct many
examples.

We shall need the following fact: if $\mathfrak{X} = (x_1, ...,$
$x_n)$ is any basis in the n-dimensional vector space
and if $\mathfrak{X}' = (y_1, ..., y_n)$ is the dual basis in $\mathfrak{V}'$,
and if the matrix of the linear transformation A in the
coordinate system $\mathfrak{X}$ is (α_{ij}) then
(2) $\alpha_{ij} = [Ax_j, y_i]$.
This follows from the definition of matrix: since
$Ax_j = \sum_k \alpha_{kj} x_k$, we have

$$[Ax_j, y_i] = \sum_k \alpha_{kj}[x_k, y_i] = \alpha_{ij}.$$

To keep things straight in the applications we rephrase
formula (2) verbally, thus: to find the (i,j) element of
[A] in the basis $\mathfrak{X}$ apply A to the j-th element of $\mathfrak{X}$
and then take the value of the i-th linear functional
(in $\mathfrak{X}'$) at the vector so obtained.

It is now very easy to find the matrix $(\alpha'_{ij}) = [A']$
in the coordinate system $\mathfrak{X}'$: we merely follow the
recipe just given. In other words we consider $A'y_j$, and

then take the value of the i-th linear functional in $\mathfrak{X}''$ (i.e. of x_i considered as a linear functional on $\mathfrak{Y}'$) at this vector; the result is that

$$\alpha'_{ij} = [x_i, A'y_j].$$

Since $[x_i, A'y_j] = [Ax_i, y_j] = \alpha_{ji}$, so that $\alpha'_{ij} = \alpha_{ji}$; this matrix $[A']$ is called the <u>transpose</u> of $[A]$.

Observe that our results on the relation between E and E' (where E is a projection) could also have been derived by using the known facts about the matricial representation of a projection together with the present result on the matrices of adjoint operators.

§34. CHANGE OF BASIS

Although what we have been doing in the preceding sections of this chapter may have been complicated it was to a large extent automatic: having introduced the new concept of linear transformation, we merely let all the preceding concepts suggest ways in which they are connected with linear transformations. We now begin the proper study of the theory of linear transformations. As a first application of this theory we shall solve the problems arising from a change of basis. These problems can be formulated without mentioning linear transformations, but their solution is most effectively given in terms of linear transformations.

Let $\mathfrak{V}$ be an n-dimensional vector space and let $\mathfrak{X} = (x_1, \ldots, x_n)$ and $\mathfrak{Y} = (y_1, \ldots, y_n)$ be two bases in $\mathfrak{V}$. We may ask the following two questions.

QUESTION I. Given a vector x in $\mathfrak{V}$, $x = \sum_i \xi_i x_i = \sum_i \eta_i y_i$, what is the relation between its coordinates $\{\xi_1, \ldots, \xi_n\}$ with respect to $\mathfrak{X}$ and its coordinates $\{\eta_1, \ldots, \eta_n\}$ with respect to $\mathfrak{Y}$?

QUESTION II. Given an ordered set of n scalars, $\{\xi_1, \ldots, \xi_n\}$, what is the relation between the vectors $x = \sum_i \xi_i x_i$ and $y = \sum_i \xi_i y_i$?

Both these questions are easily answered in the language of linear transformations. We consider, namely, the linear transformation A defined by $y_i = Ax_i$, $i = 1, \ldots, n$. More explicitly:

$$(1) \qquad A(\sum_i \xi_i x_i) = \sum_i \xi_i y_i.$$

Let (α_{ij}) be the matrix of A in the basis $\mathfrak{X}$, i.e. $y_j = Ax_j = \sum_i \alpha_{ij} x_i$. We observe that A has an inverse, since $\sum_i \xi_i y_i = 0$ implies that $\xi_1 = \xi_2 = \cdots = \xi_n = 0$.

<u>Answer to question I.</u> Since

$$\sum_j \eta_j y_j = \sum_j \eta_j A x_j = \sum_j \eta_j \sum_i \alpha_{ij} x_i$$
$$= \sum_i (\sum_j \alpha_{ij} \eta_j) x_i,$$

we have

$$(2) \qquad \xi_i = \sum_j \alpha_{ij} \eta_j.$$

<u>Answer to question II.</u>

$$(3) \qquad y = Ax.$$

Roughly speaking: the non singular linear transformation A (or, more properly, the matrix (α_{ij})) may be considered as a transformation of coordinates (as in (2)), or it may be considered (as we usually consider it, in (3)) as a transformation of vectors.

In classical treatises on vector spaces it is customary to treat vectors as numerical n-tuples rather than as abstract entities; this leads to the necessity of introducing some cumbersome terminology. We give here a brief glossary of some of the more baffling terms and notations that arise in connection with conjugate spaces and adjoint transformations.

If $\mathfrak{V}$ is an n-dimensional vector space a vector x

is given by its coordinates with respect to some preferred, absolute, coordinate system; these coordinates form an ordered set of n numbers. It is customary to write this set of n numbers in a column,

$$x = \begin{bmatrix} \xi_1 \\ \cdot \\ \cdot \\ \cdot \\ \xi_n \end{bmatrix}.$$

Elements of the conjugate space $\mathfrak{V}'$ are written as rows, $x' = [\xi_1', \ldots, \xi_n']$. If we think of x as a (rectangular) one by n matrix, and of x' as an n by one matrix, then the matrix product $x'x$ is a one by one matrix, i.e. a scalar. In our notation this number is $[x,x'] = \xi_1 \xi_1' + \cdots + \xi_n \xi_n'$. The trick of considering vectors as thin matrices works even when we consider the full grown matrices of linear transformations. Thus the matrix product of (α_{ij}) with the column (ξ_j) is the column $\eta_i = \sum_j \alpha_{ij} \xi_j$. Instead of worrying about dual bases and adjoint transformations we may form similarly the product of the row (ξ_j') with the matrix (α_{ij}) in the order $(\xi_j')(\alpha_{ij})$; the result is the row which we earlier denoted by $y' = A'x'$. The form $[Ax,x']$ is now abbreviated as $x' \cdot A \cdot x$; both dots denote ordinary matrix multiplication. The vectors x in $\mathfrak{V}$ are called <u>co-variant</u> and the vectors x' in $\mathfrak{V}'$ are called <u>contra-variant</u>. Since the notion of the product $x'x$ (i.e. $[x,x']$) depends, in this point view, on the coordinates of x and x' it becomes relevant to ask the following question: if we change basis in $\mathfrak{V}$ in accordance with the non singular linear transformation A, what must we do in $\mathfrak{V}'$ to preserve the product $x'x$? In our notation: if $[x,x'] = [y,y']$ where $y = Ax$, then how is y' related to x'? Answer: $y' = (A')^{-1}x'$. To express this whole tangle of ideas the classical terminology says that the vectors x vary <u>cogrediently</u> whereas the x' vary contragrediently.

§35. LINEAR TRANSFORMATIONS UNDER A CHANGE OF BASIS

Two questions closely related to those of the preceding section are the following.

> QUESTION III. Given a linear transformation B on $\mathfrak{V}$, what is the relation between its matrix (β_{ij}) with respect to $\mathfrak{X}$ and its matrix (γ_{ij}) with respect to $\mathfrak{Y}$?

> QUESTION IV. Given a matrix (β_{ij}) what is the relation between the linear transformations B and C defined, respectively, by $Bx_j = \sum_i \beta_{ij}x_i$ and $Cy_j = \sum_i \beta_{ij}y_i$?

Questions III and IV are explicit formulations of a problem we raised before: to one transformation there correspond (in different coordinate systems) many matrices (question III) and to one matrix there correspond many transformations (question IV).

Answer to question III. We have

(1) $$Bx_j = \sum_i \beta_{ij}x_i,$$

(2) $$By_j = \sum_i \gamma_{ij}y_i.$$

Using the linear transformation A defined in the preceding section we may write

(3) $$\begin{aligned} By_j = BAx_j &= B(\textstyle\sum_k \alpha_{kj}x_k) \\ &= \textstyle\sum_k \alpha_{kj}Bx_k = \sum_k \alpha_{kj}\sum_i \beta_{ik}x_i \\ &= \textstyle\sum_i(\sum_k \beta_{ik}\alpha_{kj})x_i, \end{aligned}$$

and

$$(4)\qquad \begin{aligned}\textstyle\sum_k \gamma_{kj} y_k &= \textstyle\sum_k \gamma_{kj} A x_k \\ &= \textstyle\sum_k \gamma_{kj} \sum_i \alpha_{ik} x_i \\ &= \textstyle\sum_i (\sum_k \alpha_{ik} \gamma_{kj}) x_i .\end{aligned}$$

Comparing (2), (3), and (4) we see that

$$(5)\qquad \textstyle\sum_k \alpha_{ik} \gamma_{kj} = \sum_k \beta_{ik} \alpha_{kj}.$$

Using matrix multiplication we write this in the dangerously simple form

$$(6)\qquad [A][C] = [B][A].$$

The danger lies in the fact that three of the four matrices written correspond to their operators in the basis $\mathfrak{X}$; the fourth one, namely the one we denoted by [C], corresponds to B in the basis $\mathfrak{Y}$. With this understanding, however, (6) is correct: a more usual form of it, adapted to computing [C] when [A] and [B] are known is

$$(7)\qquad [C] = [A]^{-1}[B][A].$$

Answer to question IV. To bring out the essentially geometric character of this question and its answer we observe that

$$(8)\qquad Cy_j = CAx_j$$

and

$$(9)\qquad \textstyle\sum_i \beta_{ij} y_i = \sum_i \beta_{ij} A x_i = A(\sum_i \beta_{ij} x_i) = ABx_j$$

Hence C is such that

$$CAx_j = ABx_j,$$

or

$$(10)\qquad CA = AB,$$

or, finally,

$$(11)\qquad C = ABA^{-1}.$$

There is no trouble with (11) similar to the reservation we had to make about the interpretation of (7): to find the operator (not matrix) C we multiply the operators

A, B, and A^{-1}, and nothing needs to be said about coordinate systems. Compare, however, the formulae (7) and (11) and observe once more the innate perversity of mathematical symbols. This is merely another aspect of the fact expressed by the formulae (1), §25, and (7), §29.

There are still too many subscripts in the answer to question IV. The validity of (11) is a geometric fact quite independent of linearity, finite dimensionality, or any other property that A, B, and C may possess: the answer to question IV is also the answer to a much more general question. This geometric question, a paraphrase of the analytic formulation of question IV, is this "If B transforms $\mathfrak{V}$, and if C transforms $A\mathfrak{V}$ the same way, what is the relation between B and C ?" The expression "the same way" is not as vague as it sounds; it means that if B takes x into, say, u, then C takes Ax into Au. The answer is, of course, the same as before: since $Bx = u$ and $Cy = v$ (where $y = Ax$ and $v = Au$), we have

$$ABx = Au = v = Cy = CAx.$$

The following is a convenient mnemonic diagram:

$$\begin{array}{ccc} & B & \\ x & \longrightarrow & u \\ A \downarrow & & \downarrow A \\ y & \longrightarrow & v \\ & C & \end{array}$$

We may go from y to v by using the short cut C, or by going around the block: in other words $C = ABA^{-1}$. Remember that ABA^{-1} is to be applied to y from right to left: first A^{-1}, then B, then A.

We have seen that the theory of changing bases is coextensive with the theory of non singular linear transformations. A non singular linear transformation is an <u>automorphism</u>, where we mean by an automorphism an isomorphism of a vector space with itself. (See §8). We observe that conversely every automorphism is a non singular linear transformation.

We hope that the relation between linear transformations and matrices is by now sufficiently clear that the reader will not object if in the sequel, when we wish to give examples of linear transformations with various properties, we content ourselves with writing down a matrix. The interpretation always to be placed on this procedure is that we have in mind the concrete vector space $\mathfrak{C}_n$ and the concrete basis $(x_1, \ldots, x_n)$ defined by $x_i = \{\delta_{i1}, \delta_{i2}, \ldots, \delta_{in}\}$. With this understanding a matrix (α_{ij}) defines, of course, a unique linear transformation A, given by the usual formula $A(\sum_i \xi_i x_i) = \sum_i (\sum_j \alpha_{ij} \xi_j) x_i$.

§36. RANGE AND NULL SPACE OF A LINEAR TRANSFORMATION

DEFINITION. If A is a linear transformation on a vector space $\mathfrak{V}$ and $\mathfrak{M}$ is any linear manifold in $\mathfrak{V}$, we denote by $A\mathfrak{M}$ the set of all vectors of the form Ax with x in $\mathfrak{M}$. The range of A is the set $\mathfrak{R}(A) = A\mathfrak{V}$; the null space of A is the set $\mathfrak{N}(A)$ of all vectors x for which $Ax = 0$.

It is immediately verified that $A\mathfrak{M}$ and $\mathfrak{N}(A)$ are linear manifolds. If we denote, as usual, by $\mathfrak{O}$ the linear manifold containing only the vector 0, it is easy to describe some familiar concepts in terms of the terminology just introduced; we list some of the results.

(i) A has an inverse if and only if $\mathfrak{R}(A) = \mathfrak{V}$ and $\mathfrak{N}(A) = \mathfrak{O}$.

(ii) In case $\mathfrak{V}$ is finite dimensional, A has an inverse if and only if $\mathfrak{N}(A) = \mathfrak{O}$

(iii) A is reduced by the linear manifold $\mathfrak{M}$ if and only if $A\mathfrak{M} \subset \mathfrak{M}$

(iv) A is completely reduced by the direct sum

decomposition $\mathfrak{V} = \mathfrak{M} \oplus \mathfrak{N}$ if and only if $A\mathfrak{M} \subset \mathfrak{M}$ and $A\mathfrak{N} \subset \mathfrak{N}$.

(v) If E is the projection on $\mathfrak{M}$ along $\mathfrak{N}$, then $\mathfrak{R}(E) = \mathfrak{M}$ and $\mathfrak{N}(E) = \mathfrak{N}$.

All these statements are easy to prove: we indicate the proof of (v). From Theorem 2, §29, we know that $\mathfrak{N}$ is the set of all solutions of the equation $Ex = 0$: this coincides with our definition of $\mathfrak{N}(E)$. We also know that $\mathfrak{M}$ is the set of all solutions of the equation $Ex = x$. If x is in $\mathfrak{M}$ then x is also in $\mathfrak{R}(E)$ since x is the E of something -- namely of x itself. Conversely if a vector x is the E of something, say $x = Ey$, (so that x is in $\mathfrak{R}(E)$), then $Ex = E^2y = Ey = x$, so that x is in $\mathfrak{M}$.

Warning: it is accidental that for projections $\mathfrak{R} \oplus \mathfrak{N} = \mathfrak{V}$. In general it need not even be true that $\mathfrak{R} = \mathfrak{R}(A)$ and $\mathfrak{N} = \mathfrak{N}(A)$ are disjoint. It can happen, for example, that for a certain vector x we have $x \neq 0$, $Ax \neq 0$, and $A^2x = 0$; for such an x, Ax clearly belongs to both the range and the null space of A. (Concrete example: A = differentiation on $\mathfrak{P}_n$, $n \geq 1$, $x = x(t) \equiv t$).

THEOREM. If A is a linear transformation on a vector space $\mathfrak{V}$ then

$$(\mathfrak{R}(A))^0 = \mathfrak{N}(A'):$$

if $\mathfrak{V}$ is finite dimensional then

$$(\mathfrak{N}(A))^0 = \mathfrak{R}(A').$$

PROOF. If y is in $(\mathfrak{R}(A))^0$ then for all x in $\mathfrak{V}$

$$0 = [Ax,y] = [x,A'y]$$

so that $A'y = 0$ and y is in $\mathfrak{N}(A')$. If, on the other hand, y is in $\mathfrak{N}(A')$ then for every x in $\mathfrak{V}$

$$0 = [x, A'y] = [Ax, y]$$

so that y is in $(\mathcal{R}(A))^{o}$.

We may apply this result to A' in place of A and we obtain

$$(1) \qquad (\mathcal{R}(A'))^{o} = \mathcal{N}(A'').$$

If $\mathfrak{V}$ is finite dimensional (and hence reflexive) we may replace, in (1), A'' by A, and then we may attach the superscript o to both sides of (1), obtaining (Theorem 2, §16)

$$\mathcal{R}(A') = (\mathcal{N}(A))^{o}$$

§37. RANK AND NULLITY

We shall now restrict attention to the finite dimensional case and draw certain easy conclusions from the theorem of the preceding section.

> DEFINITION. The <u>rank</u>, $\rho(A)$, of a linear transformation A on a finite dimensional vector space $\mathfrak{V}$ is the dimension of $\mathcal{R}(A)$; the <u>nullity</u>, $\nu(A)$, is the dimension of $\mathcal{N}(A)$.

> THEOREM 1. If A is a linear transformation on an n dimensional vector space $\mathfrak{V}$, then $\rho(A) = \rho(A')$ and $\nu(A) = n - \rho(A)$.

PROOF. The theorem of the preceding section and Theorem 1 of §16 together imply that

$$(1) \qquad \nu(A') = n - \rho(A).$$

Let $\mathfrak{X} = (x_1, \ldots, x_n)$ be any basis in $\mathfrak{V}$ for which $x_1, \ldots, x_\nu$ are in $\mathcal{N}(A)$; then for any $x = \sum_i \xi_i x_i$ we have

$$Ax = \sum_i \xi_i Ax_i = \sum_{i=\nu+1}^{n} \xi_i Ax_i.$$

In other words Ax is a linear combination of the $n-\nu$ vectors $Ax_{\nu+1}, \ldots, Ax_n$; it follows that $\rho(A) \leqq n - \nu(A)$. Applying this result to A' and using (1) we obtain

(2) $$\rho(A') \leqq n - \nu(A') = \rho(A).$$

In (2) we may replace A by A', obtaining

(3) $$\rho(A) = \rho(A'') \leqq \rho(A');$$

(2) and (3) together show that

(4) $$\rho(A) = \rho(A'),$$

and (1) and (4) together show that

(5) $$\nu(A') = n - \rho(A').$$

Replacing A by A' in (5) gives, finally,

(6) $$\nu(A) = n - \rho(A),$$

and concludes the proof of the theorem.

These results are usually discussed from a little different point of view. Let A be a linear transformation on, and $\mathfrak{X} = (x_1, \ldots, x_n)$ a basis in, the n-dimensional vector space $\mathfrak{V}$; let $[A] = (\alpha_{ij})$ be the matrix of A in the coordinate system $\mathfrak{X}$,

$$Ax_j = \sum_i \alpha_{ij} x_i.$$

Since if $x = \sum_j \xi_j x_j$, $Ax = \sum_j \xi_j Ax_j$, every vector in $\mathfrak{R}(A)$ is a linear combination of the Ax_j, and hence of any maximal linearly independent subset of the Ax_j. It follows that the maximal number of linearly independent Ax_j is precisely $\rho(A)$. In terms of the coordinates $\{\alpha_{1j}, \alpha_{2j}, \ldots, \alpha_{nj}\}$ of Ax_j we may express this by saying that $\rho(A)$ is the maximal number of linearly independent rows of the matrix $[A]$. Since (§33) the rows of $[A']$, (the matrix being expressed in terms of the dual basis of $\mathfrak{X}$) are the columns of $[A]$, it follows from Theorem 1 that $\rho(A)$ is also the maximal number of linearly independent columns of $[A]$. Hence "the row rank of $[A]$ = the column rank of $[A]$ = the rank of A."

THEOREM 2. If A is a linear transformation on the n-dimensional vector space $\mathfrak{V}$, and if $\mathfrak{Y}$ is any h-dimensional linear manifold in $\mathfrak{V}$ then the dimension of $A\mathfrak{Y}$ is $\geqq h - \nu(A)$.

PROOF. Let $\mathfrak{K}$ be any linear manifold for which $\mathfrak{V} = \mathfrak{H} \oplus \mathfrak{K}$, so that for the dimension k of $\mathfrak{K}$ we have $k = n-h$. Upon operating with A we obtain

$$A\mathfrak{V} = A\mathfrak{H} + A\mathfrak{K}$$

(The sum is not necessarily a direct sum; see §10). Since $A\mathfrak{V} = \mathfrak{R}(A)$ has dimension $n - \nu(A)$, since the dimension of $A\mathfrak{K}$ is clearly $\leqq k = n - h$, and since the dimension of the sum is $\leqq$ the sum of the dimensions, (proof ?), we have the desired result.

THEOREM 3. If A and B are linear transformations (on a finite dimensional vector space) and if B is non singular then

$$(7) \qquad \rho(AB) = \rho(BA) = \rho(A).$$

In any case

$$(8) \qquad \rho(A+B) \leqq \rho(A) + \rho(B),$$

and

$$(9) \qquad \rho(AB) \leqq \min\{\rho(A), \rho(B)\},$$

and

$$(10) \qquad \nu(AB) \leqq \nu(A) + \nu(B).$$

PROOF. Since $(AB)x = A(Bx)$, $\mathfrak{R}(AB)$ is contained in $\mathfrak{R}(A)$, so that $\rho(AB) \leqq \rho(A)$, or, in other words, the rank of a product is not greater than the rank of the first factor. Let us apply this auxiliary result to $B'A'$; this, together with what we already know, yields (9). If B has an inverse then

$$\rho(A) = \rho(AB \cdot B^{-1}) \leqq \rho(AB)$$

and

$$\rho(A) = \rho(B^{-1}BA) \leqq \rho(BA);$$

together with (9) this yields (7). (8) is an immediate consequence of an argument we have already used in the proof of Theorem 2. The proof of (10) we leave as an exercise for the reader. (Hint: apply Theorem 2 with $\mathfrak{H} = B\mathfrak{V} = \mathfrak{R}(B)$). Together the two formulae (9) and (10) are known as Sylvester's law of nullity.

§38. LINEAR TRANSFORMATIONS OF RANK ONE

We conclude our discussion of rank by a description of the matrices of transformations of rank $\leqq 1$.

THEOREM 1. If for a linear transformation A on a finite dimensional vector space $\mathfrak{V}$, $\rho(A) \leqq 1$, (i.e. $\rho(A) = 0$ or $\rho(A) = 1$), then the matrix $[A] = (\alpha_{ij})$ of A has the form $\alpha_{ij} = \beta_i \gamma_j$ in every coordinate system; conversely if the matrix of A has this form in some one coordinate system then $\rho(A) \leqq 1$.

PROOF. If $\rho(A) = 0$, $A = 0$, and the statement is trivial. If $\rho(A) = 1$, i.e. $\mathfrak{R}(A)$ is one dimensional, then there exists in $\mathfrak{R}(A)$ a non zero vector x_0 (a basis in $\mathfrak{R}(A)$) such that every vector in $\mathfrak{R}(A)$ is a multiple of x_0. Hence for every x

$$Ax = y_0 x_0,$$

where the scalar coefficient y_0 depends, of course, on $x; y_0 = y_0(x)$. The linearity of A implies that y_0 is a linear functional on $\mathfrak{V}$. Let now $\mathfrak{X} = (x_1, \ldots, x_n)$ be a basis in $\mathfrak{V}$, and let (α_{ij}) be the corresponding matrix of A, so that

$$Ax_j = \sum_i \alpha_{ij} x_i.$$

Let $\mathfrak{X}' = (y_1, \ldots, y_n)$ be the dual basis in $\mathfrak{V}'$; then (see formula (2), §33)

$$\alpha_{ij} = [Ax_j, y_i].$$

In our case

$$\alpha_{ij} = [y_o(x_j)x_o, y_i] = y_o(x_j)[x_o, y_i] = [x_o, y_i][x_j, y_o];$$

in other words we may take $\beta_i = [x_o, y_i]$ and $\gamma_j = [x_j, y_o]$.

Conversely suppose that in the fixed coordinate system $\mathfrak{X} = (x_1, \ldots, x_n)$ the matrix (α_{ij}) of A is such that $\alpha_{ij} = \beta_i \gamma_j$. We may find a linear functional $y_o = y_o(x)$ for which $\gamma_j = [x_j, y_o]$, and we may define the vector $x_o = \sum_k \beta_k x_k$. The linear transformation $\tilde{A}$ defined by $\tilde{A}x = y_o(x) \cdot x_o$ is clearly of rank one (unless, of course, $\alpha_{ij} = 0$ for all i and j), and its matrix $(\tilde{\alpha}_{ij})$, in the coordinate system $\mathfrak{X}$, is given by

$$\tilde{\alpha}_{ij} = [\tilde{A}x_j, y_i]$$

(where $\mathfrak{X}' = (y_1, \ldots, y_n)$ is the dual basis of $\mathfrak{X}$). Hence $\tilde{\alpha}_{ij} = [y_o(x_j)x_o, y_i] = [x_o, y_i][x_j, y_o] = \beta_i \gamma_j$, and since A and $\tilde{A}$ have the same matrices in one coordinate system $\tilde{A} = A$. This concludes the proof of the theorem.

The following theorem sometimes makes it possible to apply Theorem 1 to obtain results about an arbitrary linear transformation.

THEOREM 2. If A is a linear transformation, of rank ρ, on a finite dimensional vector space $\mathfrak{V}$, then A may be written as the sum of ρ transformations of rank one.

PROOF. Since $A\mathfrak{V} = \mathfrak{R}(A)$ has dimension ρ we may find ρ vectors $x_1, \ldots, x_\rho$ forming a basis for $\mathfrak{R}(A)$, so that for every vector x in $\mathfrak{V}$ we have

$$Ax = \sum_{i=1}^{\rho} \eta_i x_i,$$

where η_i depends, of course, on x; we write $\eta_i = y_i(x)$. It is easy to see that y_i is a linear functional, $y_i(x) = [x, y_i]$. In terms of these y_i we define for each $i = 1, \ldots, \rho$ a linear transformation A by $Ax = y_i(x)x_i$. Then each A_i has rank one and $A = \sum_{i=1}^{\rho} A_i$. (Compare this result with (2), §20). A slight refinement of the proof just given yields the following result.

THEOREM 3. Corresponding to any linear transformation A on a finite dimensional vector space $\mathfrak{V}$ there is a non singular linear transformation A_1 for which A_1A is a projection.

PROOF. Let $\mathfrak{R}$ and $\mathfrak{N}$ respectively be the range and the null space of A, and let $x_1, \ldots, x_\rho$ be a basis for $\mathfrak{R}$. Let $x_{\rho+1}, \ldots, x_n$ be such that $x_1, \ldots, x_n$ is a basis for $\mathfrak{V}$. Since for $i = 1, \ldots, \rho$, x_i is in $\mathfrak{R}$ we may find vectors y_i such that $Ay_i = x_i$; finally we choose a basis, which we may denote by $y_{\rho+1}, \ldots, y_n$, for $\mathfrak{N}$. We assert that $y_1, \ldots, y_n$ is a basis for $\mathfrak{V}$. We need of course prove only that the y's are linearly independent. For this purpose we suppose that $\sum_{i=1}^{n} \alpha_i y_i = 0$; then we have (remembering that for $i = \rho + 1, \ldots, n, y_i$ is in $\mathfrak{N}$)

$$A(\sum_{i=1}^{n} \alpha_i y_i) = \sum_{i=1}^{\rho} \alpha_i x_i = 0,$$

whence $\alpha_1 = \cdots \alpha_\rho = 0$. Consequently $\sum_{i=\rho+1}^{n} \alpha_i y_i = 0$; the linear independence of $y_{\rho+1}, \ldots, y_n$ shows that the remaining α's must also vanish.

A linear transformation A_1, of the kind whose existence we asserted, is now determined by the conditions $A_1 x_i = y_i$, $i = 1, \ldots n$. (For $i = 1, \ldots, \rho$, $A_1Ay_i = A_1x_i = y_i$, and for $i = \rho+1, \ldots, n$, $A_1Ay_i = A_1 0 = 0$.)

Consideration of the adjoint of A, together with the reflexivity of $\mathfrak{V}$, shows that we may also find a non singular A_2 for which AA_2 is a projection. In case A itself is non singular we must have $A_1 = A_2 = A^{-1}$.

Using the results of §35 the reader may readily verify the following matricial consequence of Theorem 3: to any matrix [A] there correspond two non singular matrices [P] and [Q] such that [P][A][Q] is a diagonal matrix whose diagonal elements are all one or zero.

§39. DETERMINANTS AND THE SPECTRAL TERMINOLOGY

It becomes necessary at this stage to go counter to the principle of giving all definitions invariantly (i.e. without using bases): we wish to say a word about determinants. At the same time we shall find it convenient to decide for once and for all what set of scalars we are going to use, and accordingly we announce that from here on throughout the remainder of this book, unless we explicitly state otherwise, we shall restrict our attention to complex vector spaces. The only special property of the field of complex numbers that we shall use in the present chapter is its algebraic closure: every polynomial equation with complex coefficients has a complex root, and, consequently, the number of its roots, counting multiplicities as usual, is exactly equal to the degree of the polynomial.

Let A be a linear transformation on an n-dimensional complex vector space $\mathfrak{V}$, and let $\mathfrak{X} = (x_1, \ldots, x_n)$ be any basis in $\mathfrak{V}$. We write $\Delta_{\mathfrak{X}}(A)$ for the determinant of the matrix $[A; \mathfrak{X}]$. We assume here a knowledge of the elementary properties of determinants; more explicitly we shall assume that the reader is aware of the following simple properties:

(1) $\Delta_{\mathfrak{X}}(0) = 0,$

(2) $\Delta_{\mathfrak{X}}(1) = 1,$

(3) $\Delta_{\mathfrak{X}}(AB) = \Delta_{\mathfrak{X}}(A) \ \Delta_{\mathfrak{X}}(B),$

(4) $\Delta_{\mathfrak{X}}(A) = \Delta_{\mathfrak{X}}(A'),$

(5) A is singular if and only if $\Delta_{\mathfrak{X}}(A) = 0,$

(6) $\Delta_{\mathfrak{X}}(A - \lambda 1)$ is a polynomial of degree n in λ; the coefficient of λ^n is $(-1)^n$.

These properties are not logically independent of each other and are not presented in the spirit of an axiomatic approach to the study of determinants: they are merely properties that we shall use. It is true, however, that if an axiomatic theory of determinants did exist, it would start with a similar list of elementary properties and then prove that there is one and only one function $\Delta_{\mathfrak{X}}$ satisfying them. We unfortunately are not able to do this without using coordinate systems, and once we resign ourselves to their use the usual explicit combinatorial definition of the determinant of a matrix is completely satisfactory for our purposes.

The most important thing to observe about $\Delta_{\mathfrak{X}}$ is that it is independent of $\mathfrak{X}$. If, in other words, $\mathfrak{X}_1$ and $\mathfrak{X}_2$ are any two bases in $\mathfrak{V}$ then $\Delta_{\mathfrak{X}_1}(B) = \Delta_{\mathfrak{X}_2}(B)$ for all B. For (2) and (3) imply that if a linear transformation A has an inverse then $\Delta_{\mathfrak{X}}(A^{-1}) = 1/\Delta_{\mathfrak{X}}(A)$; the formula (7), (§35), on change of basis, shows that

$$\Delta_{\mathfrak{X}_1}(B) = \Delta_{\mathfrak{X}_2}(A^{-1}BA)$$

for a suitable non singular A. It follows from the multiplicative property (3) that $\Delta_{\mathfrak{X}_1}(B) = \Delta_{\mathfrak{X}_2}(B)$. In view of this fact we shall in the future omit the subscript and write $\Delta(A)$ for the determinant of the matrix of A(with respect to an arbitrary coordinate system).

DEFINITION. If A is any linear transformation on a finite dimensional vector space, the <u>characteristic polynomial</u> of A is the polynomial $\Delta(A-\lambda 1)$ in λ, and the equation $\Delta(A-\lambda 1) = 0$ is the <u>characteristic equation</u> of A. A scalar λ is a <u>proper value</u>, and a vector $x \neq 0$ a <u>proper vector</u>, of A, if $Ax = \lambda x$.

Almost any combination of the adjectives proper, latent, characteristic, eigen, secular, with the nouns root, numbers, value, has been used in the literature for what we call a proper value. It is important to realize the order of choice in the definition: λ is a proper value of A if there exists a vector $x \neq 0$ for which $Ax = \lambda x$, and $x \neq 0$ is a proper vector of A if there exists a λ for which $Ax = \lambda x$. Since $Ax = \lambda x$ with an $x \neq 0$ is equivalent to $(A-\lambda 1)x = 0$, and since this in turn is equivalent to $\Delta(A-\lambda 1) = 0$, we see that λ is a proper value of A if and only if it is a root of the characteristic equation of A, and that, therefore, every A has exactly n proper values (counting multiplicities). The multiplicity of the root λ of the characteristic equation is also called the <u>multiplicity</u> of the proper value λ; in particular λ is a <u>simple</u> proper value if it is a simple root of the characteristic equation. The set of n proper values of A, with multiplicities properly counted, is the <u>spectrum</u> of A. In this language: $\lambda = 0$ is a proper value of multiplicity n of the linear transformation 0; $\lambda = 1$ is a proper value of multiplicity n of the linear transformation 1; the proper values of A, together with their multiplicities, are exactly the same as those of A'. We observe that if B is any non singular transformation then

$$\Delta(BAB^{-1} - \lambda 1) = \Delta(B(A-\lambda 1)B^{-1}) = \Delta(A-\lambda 1)$$

so that the characteristic equation, and consequently every other spectral concept such as the proper values and their multiplicities, is invariant under replacing A by BAB^{-1}.

We note also that if $Ax = \lambda x$ then

$$A^2x = A(Ax) = A(\lambda x) = \lambda(Ax) = \lambda(\lambda x) = \lambda^2 x;$$

more generally for any polynomial $p(\tau)$, $p(A)x = p(\lambda)x$, so that every proper vector x of A, belonging to the proper value λ, is also a proper vector of $p(A)$ belonging to the proper value $p(\lambda)$. Hence if A satisfies any equation of the form $p(A) = 0$ then for every proper value λ of A, $p(\lambda) = 0$. Query: what can be said about the proper values of a projection and of an involution ?

§40. MULTIPLICITIES; THE TRACE OF A LINEAR TRANSFORMATION

We call attention to another possible definition of multiplicity in order to point out and help avoid the danger of confusing the two. Suppose that λ is a proper value of A; let $\mathfrak{M}$ be the collection of all vectors x which are proper vectors of A belonging to this proper value, i.e. for which $Ax = \lambda x$. Since by our definition $x = 0$ is not a proper vector, $\mathfrak{M}$ does not contain 0; if however we enlarge $\mathfrak{M}$ so that it contains the origin then $\mathfrak{M}$ becomes a linear manifold. We might wish to define the multiplicity of λ as the dimension of this linear manifold. This is a useful concept which we shall call the <u>geometric</u> multiplicity of λ, to distinguish it from our earlier, <u>algebraic</u>, notion of multiplicity. It does not coincide with our earlier definition, as the following example shows. If D is the differentiation operator on the space $\mathfrak{P}_n$ of all polynomials of degree $\leqq n-1$, then a vector $x = x(t)$ is a proper vector of D if for some number λ, $dx/dt = \lambda x$. We borrow from the elementary theory of differential equations the fact that every solution of this equation is a constant multiple of $e^{\lambda t}$; since unless $\lambda = 0$, only the zero multiple of $e^{\lambda t}$ is a polynomial (which it must be if it is to belong to $\mathfrak{P}_n$), we must have $\lambda = 0$, and $x(t) \equiv 1$. In other words this particular operator has only one proper value (which must therefore appear with multiplicity n in the sense of our earlier definition), namely $\lambda = 0$; but, and this is more disturbing, the dimension of the linear manifold of solutions of $Ax = \lambda x$ is exactly one. Hence if $n > 1$ the two definitions of multiplicity give different values.

It is quite easy to see that the geometric multiplicity of λ is always $\leq$ its algebraic multiplicity. For if A is any linear transformation, λ_o any of its proper values, and $\mathfrak{M}$ the linear manifold of solutions of $Ax = \lambda_o x$, then it is clear that $\mathfrak{M}$ is invariant under A. If we denote by A_o the linear transformation A considered only on $\mathfrak{M}$ then it is clear (by choosing a basis in $\mathfrak{M}$, extending it to the whole space, and expressing the matrix of A in the extended basis) that $\Delta(A_o - \lambda 1)$ is a factor of $\Delta(A - \lambda 1)$. If the dimension of $\mathfrak{M}$ (= the geometric multiplicity of λ_o) is m, then $\Delta(A_o - \lambda 1) = (\lambda_o - \lambda)^m$; recalling the definition of algebraic multiplicity gives the desired result. It follows, of course, that if $\lambda_1, \ldots, \lambda_p$ are the <u>distinct</u> proper values of A, with respective geometric multiplicities $m_1, \ldots, m_p$, and if $\sum_{i=1}^{p} m_i = n$, then m_i = algebraic multiplicity of λ_i, $i=1, \ldots, p$.

Incidentally we are now able to show what we didn't quite prove before, namely that the differentiation operator on $\mathfrak{P}_n$, for $n > 1$, is not completely reducible. If it were then it could be considered as an operator on the two linear manifolds $\mathfrak{M}$ and $\mathfrak{N}$ which completely reduce it, and hence, since we know that every operator has at least one proper vector, it would have a proper vector belonging to $\mathfrak{M}$ and another one belonging to $\mathfrak{N}$.

Since this is impossible, as we showed in the preceding paragraph, the hypothesis that it is completely reducible is untenable.

By means of proper values and their multiplicities (in the sense of the algebraic definition of the preceding section) we can characterize two interesting functions of operators, one of which is the determinant and the other is something new.

Let A be any linear transformation (on an n-dimensional vector space) and let $\lambda_1, \ldots, \lambda_p$ be its different proper values. Let us denote by m_j the multiplicity of λ_j, $j=1, \ldots, p$, so that $m_1 + \cdots + m_p = n$. For any polynomial equation

$$\alpha_0 + \alpha_1 \lambda + \cdots + \alpha_n \lambda^n = 0$$

the product of the roots is $(-1)^n \alpha_0 / \alpha_n$ and the sum of the roots is $-\alpha_{n-1}/\alpha_n$. Since the leading coefficient $(= \alpha_n)$ of the characteristic equation $\Delta(A - \lambda 1) = 0$ is $(-1)^n$, and since the constant term $(= \alpha_0)$ is $\Delta(A - 0 \cdot 1) = \Delta(A)$, we have

$$(1) \qquad \Delta(A) = \prod_{j=1}^{p} \lambda_j^{m_j} .$$

This characterization of the determinant motivates the definition

$$(2) \qquad T(A) = \sum_{j=1}^{p} m_j \lambda_j :$$

$T(A)$ is called the trace of A. We shall have no occasion to use either of these numerical functions of operators in the sequel; we leave it to the interested reader to verify the following simple facts about T. If (α_{ij}) is the matrix of A in any coordinate system, then $T(A) = \sum_{i=1}^{n} \alpha_{ii}$; consequently T is a linear function of A, $T(\alpha_1 A_1 + \alpha_2 A_2) = \alpha_1 T(A_1) + \alpha_2 T(A_2)$. Also, $T(A') = T(A)$, $T(AB) = T(BA)$, $T(1) = n$, and the trace of a projection is the dimension of its range.

§41. SUPER DIAGONAL FORM

It is now trivial to prove the easiest one of the so called canonical form theorems.

THEOREM 1. If A is any linear transformation on an n-dimensional vector space $\mathfrak{V}$ then there exist $n+1$ linear manifolds $\mathfrak{M}_0, \mathfrak{M}_1, \ldots, \mathfrak{M}_{n-1}, \mathfrak{M}_n$, with the following properties:

(i) Each $\mathfrak{M}_j$, $j = 0, 1, \ldots, n-1, n$, reduces A.

(ii) The dimension of $\mathfrak{M}_j$ is j.

(iii) $(\mathfrak{O} =)\ \mathfrak{M}_0 \subset \mathfrak{M}_1 \subset \cdots\ \mathfrak{M}_{n-1} \subset \mathfrak{M}_n (= \mathfrak{V})$.

PROOF. If $n=1$ the statement is trivial; we proceed by induction, assuming that the statement is correct for $n-1$. Consider the transformation A' on $\mathfrak{V}'$; since it has at least one proper vector, say x', it is reduced by a one dimensional linear manifold $\mathfrak{M}$ -- namely the set of all multiples of x'. Let us denote by $\mathfrak{M}_{n-1}$ the annihilator (in $\mathfrak{V}'' = \mathfrak{V}$) of $\mathfrak{N}$, $\mathfrak{M}_{n-1} = \mathfrak{M}^0$; then $\mathfrak{M}_{n-1}$ is an $(n-1)$ - dimensional linear manifold in $\mathfrak{V}$, and $\mathfrak{M}_{n-1}$ reduces A. Consequently we may consider A as a linear transformation on $\mathfrak{M}_{n-1}$ alone, and we may find $\mathfrak{M}_0, \mathfrak{M}_1, \ldots, \mathfrak{M}_{n-2}, \mathfrak{M}_{n-1}$, satisfying the conditions (i), (ii), (iii). We set $\mathfrak{M}_n = \mathfrak{V}$, and we are done.

The chief interest of this theorem comes from its matricial interpretation. Since $\mathfrak{M}_1$ is one dimensional we may find in it a vector $x_1 \neq 0$. Since $\mathfrak{M}_1 \subset \mathfrak{M}_2$, x_1 is also in $\mathfrak{M}_2$, and since $\mathfrak{M}_2$ is two dimensional we may find in it a vector x_2 such that x_1 and x_2 together span $\mathfrak{M}_2$. We proceed in this way by induction, choosing vectors x_j such that $x_1, \ldots, x_j$ lie in $\mathfrak{M}_j$ and span $\mathfrak{M}_j$ for each $j=1, \ldots, n$. We obtain finally a basis $\mathfrak{X} = (x_1, \ldots, x_n)$ in $\mathfrak{V}$; let us compute the matrix of A in this coordinate system. Since x_j is in $\mathfrak{M}_j$ and since $\mathfrak{M}_j$ reduces A, Ax_j must also be in $\mathfrak{M}_j$, and is therefore a linear combination of $x_1, \ldots, x_j$. Hence in the expression

$$Ax_j = \sum_i \alpha_{ij} x_i$$

the coefficient of x_i, for $i > j$, must vanish; in other words $i > j$ implies $\alpha_{ij} = 0$. Hence the matrix of A has the superdiagonal form

$$[A] = \begin{bmatrix} \alpha_{11} & \alpha_{12} & \alpha_{13} & \ldots & \alpha_{1n} \\ 0 & \alpha_{22} & \alpha_{23} & \ldots & \alpha_{2n} \\ \ldots & \ldots & \ldots & \ldots & \ldots \\ 0 & 0 & 0 & \ldots & \alpha_{n-1,n} \\ 0 & 0 & 0 & \ldots & \alpha_{nn} \end{bmatrix}$$

It is clear from this representation that for $i=1, \ldots, n$, $\Delta(A - \alpha_{ii}1) = 0$, so that the α_{ii} are the proper values of A, appearing on the main diagonal of [A] with the proper multiplicities. We sum up:

THEOREM 2. Given any linear transformation A on an n-dimensional vector space $\mathfrak{V}$, there exists a basis $\mathfrak{X}$ in $\mathfrak{V}$ such that the matrix $[A; \mathfrak{X}]$ has the superdiagonal form; or, equivalently, given any matrix [A], there exists a non-singular matrix [B] such that $[B]^{-1}[A][B]$ is superdiagonal.

The superdiagonal form is useful for proving many results about linear transformations. It follows from it, for example, that for any polynomial $p(\tau)$ the proper values of p(A), _including multiplicities_, are precisely the numbers p(λ), where λ runs through the proper values of A.

Chapter III

ORTHOGONALITY

§42. CONCEPT OF AN INNER PRODUCT

Let us now get our feet back on the ground. We started in Chapter I by pointing out that we wish to generalize certain elementary properties of certain elementary spaces such as $\mathfrak{R}_2$. In our study so far we have done this, but we have entirely omitted from consideration one aspect of $\mathfrak{R}_2$. We have studied the qualitative aspect of linearity but we have entirely ignored the usual quantitative concepts of angle and length. In the present chapter we shall fill this gap: we shall superimpose on the vector spaces we shall study certain numerical functions corresponding to the ordinary notions of angle and length, and we shall study the new structure (vector space + given numerical function) so obtained. For a clue as to how to do this we first inspect $\mathfrak{R}_2$.

If $x = \{\xi_1, \xi_2\}$ and $y = \{\eta_1, \eta_2\}$ are any two points in $\mathfrak{R}_2$, the usual formula for the distance between x and y, or the length of the segment joining x and y, is $\sqrt{(\xi_1 - \eta_1)^2 + (\xi_2 - \eta_2)^2}$. It is convenient to introduce the notation

$$\| x \| = \sqrt{\xi_1^2 + \xi_2^2}$$

for the distance from x to the origin $0 = \{0,0\}$; in this notation the distance between x and y becomes $\| x-y \|$.

So much, for the present, for lengths and distances;

what about angles? It turns out that it is much more convenient to study, in the general case, not any of the usual measures of angles but rather their cosines. (Roughly speaking the reason for this is that the angle, in the usual picture in the circle of radius one, is the length of a certain circular arc, whereas the cosine of the angle is the length of a line segment; the latter is much easier to relate to our preceding study of linear functions). Suppose then that we let α be the angle between the segment from 0 to x and the positive ξ_1 axis, and let β be the angle between the segment from 0 to y and the same axis; the angle between the two vectors x and y is $\alpha - \beta$ so that its cosine is

$$(1) \quad \cos(\alpha-\beta) = \cos\alpha \cos\beta + \sin\alpha \sin\beta = \frac{\xi_1\eta_1 + \xi_2\eta_2}{\|x\| \cdot \|y\|}.$$

Consider the expression $\xi_1\eta_1 + \xi_2\eta_2$: by means of it we can express both angle and length by very simple formulae. We have already seen that if we know the distance between 0 and x for all x then we can compute the distance between any x and y; we assert now that if for every pair of vectors x and y we are given the value of $\xi_1\eta_1 + \xi_2\eta_2$ then in terms of this value we may compute all distances and all angles. For if we take $x = y$ then $\xi_1\eta_1 + \xi_2\eta_2$ becomes $\xi_1^2 + \xi_2^2 = \|x\|^2$, and this takes care of lengths; the formula (1), in turn, expresses the angle in terms of $\xi_1\eta_1 + \xi_2\eta_2$ and the two lengths $\|x\|$ and $\|y\|$. To have a concise notation let us write for $x = \{\xi_1, \xi_2\}$ and $y = \{\eta_1, \eta_2\}$

$$(2) \qquad \xi_1\eta_1 + \xi_2\eta_2 = (x,y);$$

what we said above is summarized by the relations

distance from 0 to $x = \|x\| = \sqrt{(x,x)}$,

distance from x to $y = \|x-y\|$,

$$\text{cosine of angle between } x \text{ and } y = \frac{(x,y)}{\|x\| \cdot \|y\|}.$$

The important properties of (x,y), considered as a numerical function of the pair of vectors x and y, are the following: it is symmetric in x and y, it depends linearly on each of its two variables, and (unless $x = 0$) (x,x) is always positive.

Observe for a moment the much more trivial picture in $\mathfrak{R}_1$. For $x = \{\xi_1\}$ and $y = \{\eta_1\}$ we should have, in this case, $(x,y) = \xi_1 \eta_1$ (and it is for this reason that (x,y) is known as the inner product or scalar product of x and y). The angle between any two vectors is either 0 or π, so that its cosine is either $+1$ or -1. This shows up the much greater sensitivity of the function (x,y) which takes on all possible numerical values.

§43. GENERALIZATION TO COMPLEX SPACES

What happens if we want to consider $\mathfrak{C}_2$ instead of $\mathfrak{R}_2$? The generalization seems to lie right at hand: for $x = \{\xi_1, \xi_2\}$ and $y = \{\eta_1, \eta_2\}$, (where now the ξ's and η's may be complex numbers), we define $(x,y) = \xi_1 \eta_1 + \xi_2 \eta_2$, and we hope that the expressions $\|x\| = \sqrt{(x,x)}$ and $\|x-y\|$ can be used as sensible measures of distance. Observe, however, the following strange phenomenon (using $i = \sqrt{-1}$):

$$\|ix\|^2 = (ix,ix) = i(x,ix) = i^2(x,x) = -\|x\|^2.$$

This means that if $\|x\|$ is positive, i.e if x is at a positive distance from the origin, then ix is not -- in fact the distance from 0 to ix is negative. This is very unpleasant: surely it is reasonable to demand that whatever it is that is going to play the role of (x,y) in this case, it should have the property that for $x=y$ it doesn't ever become negative. A formal remedy again lies close at hand: suppose that we define

$$(x,y) = \xi_1 \bar{\eta}_1 + \xi_2 \bar{\eta}_2,$$

(where the bar denotes complex conjugate). In this definition the expression (x,y) loses much of its former beauty: it is no longer quite symmetric in x and y and it is no longer quite linear in each of its variables. But, and this is what prompted us to give our new definition,

$$(x,x) = \xi_1 \bar{\xi}_1 + \xi_2 \bar{\xi}_2 = |\xi_1|^2 + |\xi_2|^2$$

is surely never negative. It is a priori dubious whether a useful and elegant theory may be built up on the basis of a function which fails to possess so many of the properties that recommended it to attention in the first place; that it is so will be shown in the sequel. A cheerful portent is this. Consider the space $\mathfrak{C}_1$ (i.e. the set of all complex numbers). It is impossible to draw a picture of any configuration in this space and then to be able to tell it apart from a configuration in $\mathfrak{R}_2$, but conceptually it is clearly a different space. The analog of (x,y) in this space, for $x = \{\xi_1\}$ and $y = \{\eta_1\}$, is $(x,y) = \xi_1 \bar{\eta}_1$, and this expression does have a simple geometric interpretation. If we join x and y to the origin by straight line segments, (x,y) will not, to be sure, be the cosine of the angle between the two segments: it turns out that for $\|x\| = \|y\| = 1$ its real part is exactly this cosine.

The complex conjugates that we were forced to introduce here will come back and plague us later: for the present we leave this heuristic introduction and turn to the formal work, after just one more comment on the notation. The similarity of the symbols $(\ ,\ )$ and $[\ ,\ ]$, the one used here for inner product and the other used earlier for linear functionals, is not accidental. We shall show later that it is, in fact, only the presence of the complex conjugation in $(\ ,\ )$ that makes it necessary to use for it a symbol different from [,], For the present, however, we cannot afford the luxury of confusing the two.

§44. FORMAL DEFINITION OF UNITARY SPACE

DEFINITION. An inner product in a (real or complex) vector space $\mathfrak{V}$ is a (respectively, real or complex) numerically valued function of the ordered pair of vectors x and y such that

(1) $(x,y) = \overline{(y,x)}$,

(2) $(\alpha_1 x_1 + \alpha_2 x_2, y) = \alpha_1(x_1,y) + \alpha_2(x_2,y)$,

(3) $(x,x) \geqq 0$; $(x,x) = 0$ is equivalent to $x = 0$.

(In the case of a real vector space the conjugation in (1) may, of course, be ignored). A unitary space is a vector space in which an inner product is defined; in a unitary space we shall use the notation $+\sqrt{(x,x)} = \|x\|$; $\|x\|$ is called the norm or length of x.

As examples of unitary spaces we may consider $\mathfrak{C}_n$, $\mathfrak{R}_n$, and $\mathfrak{P}$; in the first two cases we define, for $x = \{\xi_1, \ldots, \xi_n\}$ and $y = \{\eta_1, \ldots, \eta_n\}$, $(x,y) = \sum_i \xi_i \bar{\eta}_i$; in $\mathfrak{P}$ we define, for $x = x(t)$ and $y = y(t)$, $(x,y) = \int_0^1 x(t)\overline{y(t)}dt$.

In a unitary space we have

$$(2') \qquad (x, \beta_1 y_1 + \beta_2 y_2) = \overline{(\beta_1 y_1 + \beta_2 y_2, x)}$$

$$= \overline{\beta_1 (y_1,x) + \beta_2 (y_2,x)} = \bar{\beta}_1(x,y_1) + \bar{\beta}_2(x,y_2).$$

This fact, together with the definition of inner product, explains the terminology sometimes used to describe properties (1), (2), (3) (and their consequence (2')): (x,y) is a Hermitian symmetric (1), conjugate bilinear

((2) and (2')), positive definite (3) form. We observe that these properties of (x,y) imply for $\| x \|$ the homogeneity property

(4) $$\| \alpha x \| = | \alpha | \cdot \| x \| .$$

(Proof: $\| \alpha x \|^2 = (\alpha x, \alpha x) = \alpha\bar{\alpha}(x,x) = |\alpha|^2 \| x \|^2$).

THEOREM. (<u>Schwarz's inequality</u>). For any x and y $|(x,y)| \leq \| x \| \cdot \| y \|$; equality occurs if and only if x and y are linearly dependent.

PROOF. We write $\alpha = \| y \|^2$, $\beta = (x,y)$. Then $0 \leq \| \alpha x - \beta y \|^2 = (\alpha x - \beta y, \alpha x - \beta y) = (\alpha x, \alpha x) - (\alpha x, \beta y) - (\beta y, \alpha x) + (\beta y, \beta y)$

$$\begin{aligned} &= \| \alpha x \|^2 - [\alpha\bar{\beta}(x,y) + \bar{\alpha}\beta\overline{(x,y)}] \\ &\qquad + \| \beta y \|^2 \\ &= |\alpha|^2 \| x \|^2 - 2\, R[\alpha\beta(x,y)] \\ &\qquad + |\beta|^2 \| y \|^2 \\ &= \| y \|^4 \| x \|^2 - 2 \| y \|^2 |(x,y)|^2 \\ &\qquad + |(x,y)|^2 \| y \|^2 \\ &= \| y \|^2 (\| x \|^2 \| y \|^2 - |(x,y)|^2). \end{aligned}$$

(We use $R\zeta$ to denote the real part of the complex number ζ ; if $\zeta = \sigma + i\tau$ with real σ and τ then $R\zeta = \sigma$, and the imaginary part of ζ is $\tau = I\zeta$). If $y \neq 0$ the inequality follows by dividing out $\| y \|^2$; if $y = 0$ we clearly have equality. More generally since the only place an inequality came in the above computation was in the assertion $\| \alpha x - \beta y \|^2 \geq 0$, it is clear that the vanishing of the last term implies the linear dependence of x and y. The reader may easily verify the converse.

§45. APPLICATIONS OF SCHWARZ'S INEQUALITY

The Schwarz inequality has important arithmetic, geometric, and analytic consequences.

(1) In any unitary space we define the distance $\delta(x,y)$ between two vectors x and y by

$$\delta(x,y) = \| x-y \| = \sqrt{(x-y,\ x-y)}.$$

In order for δ to deserve to be called a distance it should have the following three properties:

(i) $\delta(x,y) = \delta(y,x)$;

(ii) $\delta(x,y) \geq 0$; $\delta(x,y) = 0$ is equivalent to $x=y$;

(iii) $\delta(x,y) \leq \delta(x,z) + \delta(z,y)$.

(In a vector space it is also pleasant to be sure that distance is invariant under translations:

(iv) $\delta(x,y) = \delta(x+z,\ y+z)$.)

Properties (i), (ii), and (iv) are obviously possessed by the particular δ we defined; the only question is concerning the "triangle inequality" (iii). To prove the validity of (iii) we observe that

$$\begin{aligned} \| x+y \|^2 &= (x+y,\ x+y) = \| x \|^2 + 2R(x,y) + \| y \|^2 \\ &\leq \| x \|^2 + 2|(x,y)| + \| y \|^2 \\ &\leq \| x \|^2 + 2\| x \| \cdot \| y \| + \| y \|^2 \\ &= (\| x \| + \| y \|)^2; \end{aligned}$$

replacing x by $x-z$ and y by $z-y$ we obtain

$$\| x-y \| \leq \| x-z \| + \| x-y \|,$$

and this is equivalent to (iii).

(2) In the space $\mathfrak{R}_n$, $\dfrac{(x,y)}{\| x \| \cdot \| y \|}$ is the cosine of the angle between x and y. The Schwarz inequality in this case merely amounts to the statement that the cosine of a real angle is ≤ 1.

(3) In the space $\mathfrak{C}_n$ the Schwarz inequality becomes the so-called Cauchy inequality: for any two sets $\{\xi_1, \ldots, \xi_n\}$ and $\{\eta_1, \ldots, \eta_n\}$ of complex numbers we have

$$|\textstyle\sum_{i=1}^n \xi_i \bar{\eta}_i|^2 \leq \sum_{i=1}^n |\xi_i|^2 \cdot \sum_{i=1}^n |\eta_i|^2.$$

(4) In the space $\mathfrak{P}$ the Schwarz inequality becomes

$$|\textstyle\int_0^1 x(t)\overline{y(t)}\,dt|^2 \leq \int_0^1 |x(t)|^2\,dt \cdot \int_0^1 |y(t)|^2 dt.$$

It is useful to observe that the relations mentioned in (1)-(4) above are not only analogous to the general Schwarz inequality, but actually consequences or special cases of it.

(5) We mention in passing that there is room between the two notions (general vector spaces and unitary spaces) for an intermediate concept that has been studied extensively in recent years. This concept is that of a normed vector space, i.e. a vector space in which there is an acceptable definition of length, but nothing is said about angles. A norm in a vector space is a numerically valued function $\| x \|$ of the vectors x such that $\| x \| > 0$ unless $x = 0$, $\| \alpha x \| = | \alpha | \; \| x \|$, and $\| x+y \| \leq \| x \| + \| y \|$. Our discussion so far shows that a unitary space is a normed vector space; the converse is not in general true. In other words if all we are given is $\| x \|$ satisfying the three conditions just given, it may not be possible to find an inner product (x,y) for which $(x,x) = \| x \|^2$. The norm in a unitary space has an essentially quadratic character; it can be shown for example that a necessary and sufficient condition for the existence of an inner product giving rise to a preassigned norm $\| x \|$ is the general validity of the relation $\| x+y \|^2 + \| x-y \|^2 = 2(\| x \|^2 + \| y \|^2)$. (Compare formula (4), §59.)

§46. ORTHOGONALITY

The most important relation between vectors of a unitary space is orthogonality.

DEFINITION. x and y are orthogonal if $(x,y) = 0$. (Observe that the relation is symmetric: since $(x,y) = \overline{(y,x)}$, (x,y) and (y,x) vanish together).

If we recall the motivation for the introduction of

(x,y), the terminology explains itself: the two vectors are orthogonal (or perpendicular) if the cosine of the angle between them is 0, so that the angle between them is 90°.

Two linear manifolds are orthogonal if every vector in each is orthogonal to every vector in the other. A set $\mathfrak{X}$ of vectors is an orthonormal set if for every x and y both in $\mathfrak{X}$ we have $(x,y) = 0$ or $(x,y) = 1$ according as $x \neq y$ or $x = y$. (If $\mathfrak{X}$ is finite, $\mathfrak{X} = (x_1, \ldots, x_n)$, we have $(x_i, x_j) = \delta_{ij}$).

To make our last definition in this connection we first observe that an orthonormal set is linearly independent. For if $x_1, \ldots, x_k$ is any finite subset of the orthonormal set $\mathfrak{X}$, then $\sum_i \alpha_i x_i = 0$ implies

$$0 = (\sum_i \alpha_i x_i, x_j) = \sum_i \alpha_i (x_i, x_j) = \sum_i \alpha_i \delta_{ij} = \alpha_j:$$

in other words a linear combination of x's can vanish only if all the coefficients vanish. Hence: in a finite dimensional unitary space the number of vectors in an orthonormal set is always finite and, in fact, not greater than the linear dimension of the space. We define, in this case, the orthogonal dimension of the space, as the largest number of vectors an orthonormal set can contain. We call an orthonormal set complete if it is not contained in any larger orthonormal set.

(Warning: For all we know at this stage the concepts of orthogonality and orthonormal sets are vacuous. It is easy to see, however, that if the space contains a vector $x \neq 0$, then we can always find orthogonal vectors and orthonormal sets; for example x and 0 are orthogonal, and the set consisting of $x/\| x \|$ alone is an orthonormal set. We grant that the example of orthogonal vectors we just gave is not much more inspiring than the example $x = 0$, $y = 0$, but we shall show presently that there are always "enough" orthogonal vectors to operate with in comfort. Observe also that we have no right to assume that the number of elements in a complete orthonormal set is the orthogonal dimension. The point is this: if we had an orthonormal set with that many elements, it would clearly be complete; but it is conceivable that some other set contains fewer elements, but is still complete because its nasty structure precludes the possibility of extending it. These difficulties are purely verbal and will evaporate the moment we start proving things: they occur only because from among the several possibilities for the definition of completeness we had to choose a definite one and we must prove its equivalence with the others).

We need some notation. If $\mathfrak{C}$ is any set of vectors of a unitary space $\mathfrak{V}$, we denote by $\mathfrak{C}^{\perp}$ the set of all vectors orthogonal to every vector in $\mathfrak{C}$; it is clear that $\mathfrak{C}^{\perp}$ is a linear manifold (whether or not $\mathfrak{C}$ is one), and that $\mathfrak{C}$ is contained in $\mathfrak{C}^{\perp\perp} = (\mathfrak{C}^{\perp})^{\perp}$. It follows that the linear manifold spanned by $\mathfrak{C}$ is contained in $\mathfrak{C}^{\perp\perp}$. In case $\mathfrak{C}$ is a linear manifold we shall call $\mathfrak{C}^{\perp}$ the <u>orthogonal complement</u> of $\mathfrak{C}$. We use the sign $\perp$ in order to be reminded of orthogonality (or perpendicularity). $\mathfrak{C}^{\perp}$ might be pronounced as "C perp."

§47. CHARACTERIZATIONS OF COMPLETENESS

THEOREM 1. (<u>Bessel's inequality</u>). If $\mathfrak{X} = (x_1, \ldots, x_n)$ is any finite orthonormal set in a unitary space, and x is any vector, and if we write $\alpha_i = (x, x_i)$, then

$$\sum_i |\alpha_i|^2 \leq \| x \|^2.$$

Moreover $x' = x - \sum_i \alpha_i x_i$ is orthogonal to each x_j and consequently to the linear manifold spanned by $\mathfrak{X}$.

PROOF. We have

$$0 \leq \| x' \|^2 = (x',x') = (x-\sum_i \alpha_i x_i,\ x-\sum_j \alpha_j x_j)$$
$$= (x,x) - \sum_i \alpha_i (x_i,x) - \sum_j \overline{\alpha}_j (x,x_j) +$$
$$+ \sum_i \sum_j \alpha_i \overline{\alpha}_j (x_i,x_j)$$
$$= \| x \|^2 - \sum_i | \alpha_i |^2 - \sum_i | \alpha_i |^2 + \sum_i | \alpha_i |^2$$
$$= \| x \|^2 - \sum_i | \alpha_i |^2$$

and

$$(x',x_j) = (x,x_j) - \sum_i \alpha_i (x_i,x_j) = \alpha_j - \alpha_j = 0.$$

THEOREM 2. Let $\mathfrak{X} = (x_1, \ldots, x_n)$ be any finite orthonormal set in a unitary space $\mathfrak{V}$; the following six conditions on $\mathfrak{X}$ are equivalent to each other.

(1) $\mathfrak{X}$ is complete.

(2) $(x,x_i) = 0$ for $i = 1, \ldots, n$ implies $x = 0$.

(3) The linear manifold spanned by $\mathfrak{X}$ is the whole space $\mathfrak{V}$.

(4) For every x in $\mathfrak{V}$, $x = \sum_i (x,x_i)x_i$.

(5) For every pair x,y in $\mathfrak{V}$, $(x,y) = \sum_i (x,x_i)(x_i,y)$. (Parseval's identity).

(6) For every x in $\mathfrak{V}$, $\| x \|^2 = \sum_i |(x,x_i)|^2$.

PROOF. We shall establish the implications (1) $\Longrightarrow$ (2) $\Longrightarrow$ (3) $\Longrightarrow$ (4) $\Longrightarrow$ (5) $\Longrightarrow$ (6) $\Longrightarrow$ (1). Thus we first assume (1) and prove (2), then assume (2) to prove (3), and so on until finally we prove (1) assuming (6).

(1) $\Longrightarrow$ (2). If $(x,x_i) = 0$ for all i and $x \neq 0$ then we may adjoin $x/\| x \|$ to $\mathfrak{X}$ and thus obtain an orthonormal set larger than $\mathfrak{X}$.

(2) $\Longrightarrow$ (3). If there is an x which is not a linear combination of the x_i then, by the last part of Theorem 1, $x' = x - \sum_i (x,x_i)x_i$ is different from 0 and is

orthogonal to each x_i.

(3) $\Longrightarrow$ (4). We know that every x has the form $x = \sum_j \alpha_j x_j$; it follows that $(x, x_i) = \sum_j \alpha_j (x_j, x_i) = \alpha_i$.

(4) $\Longrightarrow$ (5). We have $x = \sum_i \alpha_i x_i$, $y = \sum_j \beta_j x_j$, with $\alpha_i = (x, x_i)$, $\beta_j = (y, x_j)$. It follows that $(x,y) = (\sum_i \alpha_i x_i, \sum_j \beta_j x_j) = \sum_i \sum_j \alpha_i \bar{\beta}_j (x_i x_j) = \sum_i \alpha_i \bar{\beta}_i$.

(5) $\Longrightarrow$ (6). Set $x = y$.

(6) $\Longrightarrow$ (7). If $\mathfrak{X}$ were contained in a larger read (1) orthonormal set, say if x_0 is orthogonal to each x_i, then $\| x_0 \|^2 = \sum_i \|(x_0, x_i)\|^2 = 0$ so that $x_0 = 0$.

§48. EXISTENCE OF COMPLETE ORTHONORMAL SETS

THEOREM. Let $\mathfrak{V}$ be an n-dimensional unitary space. Then complete orthonormal sets in $\mathfrak{V}$ exist and every complete orthonormal set contains exactly n elements, so that the orthogonal dimension of $\mathfrak{V}$ is the same as its linear dimension.

PROOF. To people not fussy about hunting for an element of a possibly uncountable set, the existence is obvious. We have already seen that orthonormal sets exist, so we choose one; if it is not complete we may increase it, and if the resulting orthonormal set is still not complete we increase it again, and we proceed in this way by induction. Since an orthonormal set may contain at most n elements, in at most n steps we will have reached a complete orthonormal set. This set spans the whole space (§47, Theorem 2, (1) $\Longrightarrow$ (3)), and since it is also linearly independent, it is a basis and hence contains precisely n elements.

There is a constructive method of avoiding this crude induction, and since it sheds further light on the notions involved we reproduce it here as an alternative proof of the theorem.

The Gram-Schmidt orthogonalization process. Let $\mathfrak{X} = (x_1, \ldots, x_n)$ be any basis in $\mathfrak{V}$. We shall construct a complete orthonormal set $\mathfrak{Y} = (y_1, \ldots, y_n)$ with the property that each y_j is a linear combination of $x_1, \ldots, x_j$. To do this for $j = 1$, we need only to observe that $x_1 \neq 0$ (since $\mathfrak{X}$ is linearly independent); we write $y_1 = x_1/\| x_1 \|$. Suppose then that $y_1, \ldots y_r$ have been found so that they form an orthonormal set and so that each $y_j (j = 1, \ldots, r)$ is a linear combination of $x_1, \ldots, x_j$. We write $z = x_{r+1} - (\alpha_1 y_1 + \cdots + \alpha_r y_r)$, where $\alpha_1, \ldots, \alpha_r$ are any scalars, and we observe that for $j = 1, \ldots, r$

$$(z, y_j) = (x_{r+1} - \sum_i \alpha_i y_i, y_j) = (x_{r+1}, y_j) - \alpha_j,$$

so that if we choose $\alpha_j = (x_{r+1}, y_j)$, then $(z, y_j) = 0$ for $j = 1, \ldots, r$. Since, moreover, z is a linear combination of x_{r+1} and $y_1, \ldots, y_r$, it is also a linear combination of x_{r+1} and $x_1, \ldots, x_r$. Finally z is different from zero, since $x_1, \ldots, x_r, x_{r+1}$ are linearly independent and the coefficient of x_{r+1} in the expression for z is not zero. We write $y_{r+1} = z/\| z \|$; clearly $y_1, \ldots, y_r, y_{r+1}$ is again an orthonormal set with all the desired properties, and the induction step is established. We shall make use of the fact that not only is y_j a linear combination of the first j x's but conversely each x_j is a linear combination of the first j y's.

We shall find it convenient and natural, in unitary spaces, to work exclusively with such bases as are also complete orthonormal sets. We shall call such a basis an <u>orthogonal basis</u> or an <u>orthogonal coordinate system</u>; whenever we shall discuss not necessarily orthogonal bases we shall emphasize this fact by calling them linear bases.

§49. PROJECTION THEOREM

Since a linear manifold in a unitary space may itself be considered as a unitary space, the theorem of the preceding section may be applied. As the most important application of this fact we have the following projection theorem.

THEOREM. If $\mathfrak{M}$ is any linear manifold in a finite dimensional unitary space $\mathfrak{V}$, then $\mathfrak{V}$ is the direct sum of $\mathfrak{M}$ and $\mathfrak{M}^{\perp}$, and $\mathfrak{M}^{\perp\perp} = \mathfrak{M}$.

PROOF. Let $\mathfrak{X} = (x_1, \ldots, x_m)$ be an orthonormal set which is complete in $\mathfrak{M}$, and let z be any vector in $\mathfrak{V}$. We write $x = \sum_{i=1}^{m} \alpha_i x_i$, where $\alpha_i = (z, x_i)$; it follows from §47, Theorem 1, that $y = z - x$ is in $\mathfrak{M}^{\perp}$, so that z may be written as a sum of two vectors, $z = x + y$, with x in $\mathfrak{M}$ and y in $\mathfrak{M}^{\perp}$. That $\mathfrak{M}$ and $\mathfrak{M}^{\perp}$ are disjoint is clear: if x belonged to both then we should have $\| x \|^2 = (x,x) = 0$. It follows from the theorem of §17 that $\mathfrak{V} = \mathfrak{M} \oplus \mathfrak{M}^{\perp}$

We observe that in the decomposition $z = x + y$, we have $(z,x) = (x+y,x) = \| x \|^2 + (y,x) = \| x \|^2$, and similarly $(z,y) = \| y \|^2$. Hence, if z is in $\mathfrak{M}^{\perp\perp}$, so that $(z,y) = 0$, then $\| y \|^2 = 0$, so that $z = x$ is in $\mathfrak{M}$: in other words $\mathfrak{M}^{\perp\perp}$ is contained in $\mathfrak{M}$. Since we already know that $\mathfrak{M}$ is contained in $\mathfrak{M}^{\perp\perp}$, the proof of the theorem is complete.

This kind of direct sum decomposition of a unitary space (i.e. by a manifold and its orthogonal complement) is of considerable geometric interest. We shall study a little later the associated projections: they turn out to be an interesting and important subclass of the class of all projections. At present we remark only on the connection with the Pythagorean theorem: since $(z,x) = \| x \|^2$ and $(z,y) = \| y \|^2$, we have

$$\| z \|^2 = (z,z) = (z,x) + (z,y) = \| x \|^2 + \| y \|^2.$$

In other words the square of the hypotenuse is the sum of the squares of the sides. More generally if $\mathfrak{M}_1$, ..., $\mathfrak{M}_k$ is any collection of pairwise orthogonal linear manifolds in a unitary space $\mathfrak{V}$, and if $x = x_1 + \cdots + x_k$, with x_j in $\mathfrak{M}_j$ for $j = 1, \ldots, k$, then

$$\| x \|^2 = \| x_1 \|^2 + \cdots + \| x_k \|^2.$$

§50. REPRESENTATION OF LINEAR FUNCTIONALS

We are now in a position to study linear functionals on unitary spaces. For a general n-dimensional vector space, the conjugate space is also n-dimensional and is therefore isomorphic to the original space. There is, however, no obvious natural isomorphism that we can set up -- we have to wait for the second conjugate space to get back where we came from. The main point of the theorem we shall prove is that in unitary spaces there is a "natural" correspondence between $\mathfrak{V}$ and $\mathfrak{V}'$: the only cloud on the horizon is that it is not quite an isomorphism.

THEOREM. To any linear functional $y' = y'(x)$ on a finite dimensional unitary space $\mathfrak{V}$ there corresponds a unique vector y in $\mathfrak{V}$ such that, for all x, $y'(x) = (x,y)$.

PROOF. If $y' = 0$ we may choose $y = 0$; let us from now on assume that y' is not identically zero. Let $\mathfrak{M}$ be the linear manifold of all vectors x for which $y'(x) = 0$, and let $\mathfrak{N} = \mathfrak{M}^{\perp}$ be the orthogonal complement of $\mathfrak{M}$. Then $\mathfrak{N}$ contains a vector $y_0 \neq 0$; by multiplication with a suitable constant we may assume $\| y_0 \| = 1$. We write $y = \overline{y'(y_0)} \cdot y_0$ (where the bar denotes, as usual, complex conjugation); we do then have the desired relation

(1) $\qquad y'(x) = (x,y)$

at least for $x = y_0$ and for all x in $\mathfrak{M}$. For an arbitrary x in $\mathfrak{B}$ we write $x_0 = x - \lambda y_0$ where

$$\lambda = \frac{y'(x)}{y'(y_0)};$$

then $y'(x_0) = 0$ and $x = x_0 + \lambda y_0$ is a linear combination of two vectors for each of which (1) is valid. From the linearity of both sides of (1) it follows that (1) holds for x, as was to be proved.

To prove uniqueness suppose that $(x,y_1) = (x,y_2)$ for all x; then $(x,y_1-y_2) = 0$ for all x and therefore in particular for $x = y_1-y_2$, so that $\|y_1-y_2\|^2 = 0$, and $y_1 = y_2$.

The correspondence $y' \rightleftarrows y$ is a one to one correspondence between $\mathfrak{B}$ and $\mathfrak{B}'$, with the property that to $y_1'+y_2'$ there corresponds y_1+y_2 and to $\alpha y'$ there corresponds $\bar{\alpha} y$: for this reason we refer to it as a <u>conjugate isomorphism</u>. In spite of the fact that this conjugate isomorphism makes $\mathfrak{B}'$ practically indistinguishable from $\mathfrak{B}$, it is wise to keep the two conceptually separate. One reason for this is that we should like $\mathfrak{B}'$ to be a unitary space along with $\mathfrak{B}$; if, however, we follow the clue given by the conjugate isomorphism between $\mathfrak{B}$ and $\mathfrak{B}'$ the conjugation again causes trouble. Let y_1' and y_2' be any two elements of $\mathfrak{B}$; if $y_1'(x) = (x,y_1)$ and $y_2'(x) = (x,y_2)$, the temptation is great to define

$$(y_1',y_2') = (y_1,y_2).$$

A moment's consideration will show that this expression does not satisfy (2) (§44) and is not therefore a suitable inner product: thus we have

$$(\alpha y_1',y_2') = (\bar{\alpha} y_1,y_2) = \bar{\alpha}(y_1,y_2) = \bar{\alpha}(y_1',y_2').$$

The remedy is clear: we define

$$(y_1', y_2') = \overline{(y_1, y_2)} = (y_2, y_1);$$

we leave it to the reader to verify that with this definition $\mathfrak{V}'$ becomes a unitary space. We shall denote this unitary space by $\mathfrak{V}^*$.

§51. RELATION BETWEEN PARENTHESES AND BRACKETS

It becomes necessary now to straighten out the relation between general vector spaces and unitary spaces. The theorem of the preceding section shows that, as long as we are careful about complex conjugation, (x,y) can completely take the place of $[x,y]$. It might seem that it would have been desirable to develop the entire subject of general vector spaces in such a way that the concept of orthogonality in a complex unitary becomes not merely an analog but a special case of some previously studied general relation between vectors and functionals. One way, for example, of avoiding the unpleasantness of conjugation (or, rather, of shifting it to a less conspicuous position) would have been to define the conjugate space of a complex vector space as the set of conjugate linear functionals, i.e. the set of numerical valued functions $y(x)$ for which

$$y(\alpha_1 x_1 + \alpha_2 x_2) = \bar{\alpha}_1 y(x_1) + \bar{\alpha}_2 y(x_2).$$

Because it seemed pointless (and contrary to common usage) to introduce this complication into the general theory we chose instead the roundabout way that we just travelled. Since from now on we shall deal with unitary spaces only, we ask the reader mentally to revise all the preceding work by replacing, throughout, the bracket $[x,y]$ by the parenthesis (x,y). Let us examine the effect of this change on the theorems and definitions of the first two chapters.

Replacing $\mathfrak{V}'$ by $\mathfrak{V}^*$ is merely a change of notation: the new symbol is supposed to remind us that something new (namely an inner product) has been added to $\mathfrak{V}'$. Of a little more interest is the (conjugate) isomorphism between $\mathfrak{V}$ and $\mathfrak{V}^*$: by means of it the theorems of §14, asserting the existence of linear functionals with various properties, may now be interpreted as asserting the existence of certain vectors in $\mathfrak{V}$ itself. Thus, for example, the existence of a dual basis to any given basis $\mathfrak{X} = (x_1, \ldots, x_n)$ implies now the existence of a basis $\mathfrak{Y} = (y_1, \ldots, y_n)$ (of $\mathfrak{V}$) with the property that $(x_i, y_j) = \delta_{ij}$. Query: what does it mean for a basis to be <u>self-dual</u>, i.e. $x_i = y_i$, $i = 1, \ldots n$?

More exciting still is the implied replacement of the annihilator $\mathfrak{M}^0$ of a linear manifold $\mathfrak{M}$, ($\mathfrak{M}^0$ lying in $\mathfrak{V}'$ or $\mathfrak{V}^*$) by the orthogonal complement $\mathfrak{M}^\perp$ (lying, along with $\mathfrak{M}$, in $\mathfrak{V}$). The most radical new development, however, concerns the adjoint of a linear transformation. Thus we may write the analog of (1) (§32) and corresponding to every linear transformation A on $\mathfrak{V}$ we may define a linear transformation A* by the relation

(1) $(Ax, y) = (x, A^*y)$.

A* is again a linear transformation defined on the same vector space $\mathfrak{V}$, but because of the Hermitian symmetry of (x,y) the relation between A and A* is not quite the same as the relation between A and A': the most notable difference is that $(\alpha A)^* = \bar{\alpha} A^*$ (and <u>not</u> $= \alpha A^*$). Associated with this phenomenon is the fact that if the matrix of A, with respect to some fixed basis, is (α_{ij}) then the matrix of A*, with respect to the dual basis, is not (α_{ji}) but $(\overline{\alpha_{ji}})$; also for determinants we do not have $\Delta(A^*) = \Delta(A)$ but $\Delta(A^*) = \overline{\Delta(A)}$, and, consequently, the proper values of A* are not the same as those of A, but rather their conjugates. Here, however, the differences stop. All the other results of §32 on the anti-isomorphic nature of the correspondence $A \rightleftarrows A^*$ are valid;

the identity $A = A^{**}$ is strictly true and does not need the help of an isomorphism to interpret it.

Presently we shall discuss linear transformations on unitary spaces and we shall see that the principal new feature differentiating their study from the discussion of Chapter II is the possibility of comparing A and A^* as linear transformations on the same space, and of investigating those classes of linear transformations which bear a particularly simple relation to their adjoints.

§52. COMPARISON OF THE TWO "NATURAL" ISOMORPHISMS FROM $\mathfrak{V}$ TO $\mathfrak{V}^{**}$

There is now only one more possible doubt that the reader might (or, at any rate, should) have. Many of our preceding results were consequences of such reflexivity relations as $A^{**} = A$; do these remain valid after the brackets to parentheses revolution? More to the point is the following way of asking the question: everything we say about the unitary space $\mathfrak{V}$ must also be true about the unitary space $\mathfrak{V}^*$; in particular it is also in a natural conjugate isomorphic relation with its conjugate space $\mathfrak{V}^{**}$. If now to every vector in $\mathfrak{V}$ we make correspond a vector in $\mathfrak{V}^{**}$, by first applying the natural conjugate isomorphism from $\mathfrak{V}$ to $\mathfrak{V}^*$ and then going the same way from $\mathfrak{V}^*$ to $\mathfrak{V}^{**}$, then this mapping is a rival for the title of natural mapping from $\mathfrak{V}$ to $\mathfrak{V}^{**}$ -- a title already awarded in Chapter I to a seemingly different correspondence. What is the relation between the two natural correspondences? Our statements about the coincidence, except for trivial modifications, of the parenthesis and bracket theories, are really justified by the fact, which we shall now show, that the two mappings are the same. (It should not be surprising, since $\bar{\bar{\alpha}} = \alpha$, that after two applications the bothersome conjugation disappears). The proof is shorter than the introduction to it.

Let y_o be any element of $\mathfrak{V}$; to it there corresponds the linear functional $y_o^* = y_o^*(x) = (x, y_o)$ in $\mathfrak{V}^*$, and to y_o^* in turn there corresponds the linear functional $y_o^{**} = y_o^{**}(y^*) = (y, y_o^*)$ in $\mathfrak{V}^{**}$. Both these correspondences are given by the mapping introduced in this chapter. Previously (see §15) the correspondent y_o^{**} in $\mathfrak{V}^{**}$ of y_o in $\mathfrak{V}$ was defined by $y_o^{**}(y^*) = y^*(y_o)$ for all y^* in $\mathfrak{V}^*$: we must show that y_o^{**}, as we here defined it, satisfies this relation. Let $y^* = y^*(x) = (x, y)$ be any linear functional on $\mathfrak{V}$, (i.e. any element of $\mathfrak{V}^*$); we have

$$y_o^{**}(y^*) = (y^*, y_o^*) = (y_o, y) = y^*(y_o).$$

(The middle equality comes from the definition of inner product in $\mathfrak{V}^*$). This settles all our problems.

A word about direct sums. We may define the direct sum of two unitary spaces $\mathfrak{U}$ and $\mathfrak{V}$ just as we defined (in §17) the direct sum of any two vector spaces: it is only necessary to say something about the inner product. The obvious solution works in this case: we define (using the notation of §17)

$$(\{x_1, y_1\} \ \{x_2, y_2\}) = (x_1, x_2) + (y_1, y_2).$$

There is not much more to be said. We should prove the analog of the theorem of §17, i.e. we should be able to decide when a unitary space may be considered as the direct sum of two of its subspaces. We leave it to the reader to verify that a necessary and sufficient condition is that the two linear manifolds be orthogonal complements of each other.

§53. LINEAR TRANSFORMATIONS ON A UNITARY SPACE

Let us now study the algebraic structure of the class of all linear transformations on a unitary space $\mathfrak{V}$. In many fundamental respects this class resembles the class of all complex numbers. In both systems the notions of addition, multiplication, 0, and 1 are defined and have similar properties, and in both systems there is an involutory (anti-) automorphism of the system on itself -- namely $A \longrightarrow A^*$ and $\zeta \longrightarrow \bar{\zeta}$. We shall use this analogy as a heuristic principle and we shall attempt to carry over to linear transformations some well known concepts of the complex domain. We will be hindered in this work by two properties of linear transformations, of which, possibly surprisingly, the second is much more serious: the impossibility of unrestricted division and the non commutativity of general linear transformations.

First we need an auxiliary result.

THEOREM. If A is a linear transformation on a complex unitary space $\mathfrak{V}$ then the vanishing, identically, of either of the two expressions (Ax,x) and (Ax,y) is necessary and sufficient for A to be zero.

PROOF. That either condition is necessary is clear, as also is the fact that the second condition is sufficient. (If $(Ax,y) = 0$ for all x and y then choose $y = Ax$; it follows that $Ax = 0$ for all x). In fact this sufficiency has its analog in pure vector space theory: $[Ax,y]$ vanishes identically if and only if $A = 0$. (If $[Ax,y] = 0$ for all y, then $Ax = 0$ for each x, and hence $A = 0$). It is the sufficiency of the condition $(Ax,x) = 0$ that is really peculiar to complex unitary spaces.

For the proof of this sufficiency we use the so-called <u>polarization</u> identity:

$$(1)\quad \alpha\bar{\beta}(Ax,y) + \bar{\alpha}\beta(Ay,x) = (A(\alpha x+\beta y), (\alpha x+\beta y)) - |\alpha|^2(Ax,x) - |\beta|^2(Ay,y).$$

(We leave to the reader the simple verification carried out by expanding the first term on the right). If (Ax,x)

is identically zero then we obtain, first choosing $\alpha = \beta = 1$, and then $\alpha = i = \sqrt{-1}$, $\beta = 1$,

$$(Ax,y) + (Ay,x) = 0$$

$$i(Ax,y) - i(Ay,x) = 0.$$

Dividing the second of these two equations by i and then forming their arithmetic mean we see that $(Ax,y) = 0$ for all x and y, so that, by the easier half of our theorem, $A = 0$.

This process of polarization is often used to get information about the "bilinear form" (Ax,y) when only knowledge of the "quadratic form" (Ax,x) is assumed.

It is important to observe that this seemingly innocuous auxiliary theorem uses very essentially the complex number system; it and many of its consequences fail to be true in a real vector space. The proof of course breaks down at our choice of $\alpha = \sqrt{-1}$. For an example: a 90° rotation of the plane clearly has the property that it sends every vector x into a vector Ax orthogonal to it.

As a curiousity concerning the form (Ax,x) we mention the following fact. The set of all possible values of (Ax,x), as x ranges over the unit sphere (i.e. the set of all vectors x for which $\|x\| = 1$) is a convex set (in the complex plane) containing all proper values of A. For a special kind of transformations (namely normal transformations; see §64) this convex set is the smallest convex polygon determined by the proper values of A.

§54. HERMITIAN TRANSFORMATIONS

The three most important subsets of the complex number plane are the real numbers, the positive real numbers, and the numbers of absolute value one. We shall now proceed systematically to use our heuristic analogy of transformations with complex numbers and try to discover the analogs among transformations of these well known numerical concepts.

When is a complex number real? Clearly a necessary and sufficient condition for the reality of ζ is the equation $\zeta = \bar{\zeta}$. We might accordingly (remembering that the analog of the complex conjugate for linear transformations is the adjoint) define a linear transformation A to be real if $A = A^*$. More commonly linear transformations A for which $A = A^*$ are called Hermitian (or symmetric, Hermitian symmetric, self-adjoint). We shall see that Hermitian transformations do indeed play the same role as real numbers: the first indication that they are tied up with the concept of reality in more ways than through the formal analogy that suggested their definition is the following theorem.

THEOREM. A necessary and sufficient condition that a linear transformation A defined on a complex unitary space be Hermitian is that (Ax,x) be real for all x.

PROOF. If $A = A^*$ then

$$(Ax,x) = (x,A^*x) = (x,Ax) = \overline{(Ax,x)},$$

so that (Ax,x) is equal to its own conjugate and is therefore real. If conversely (Ax,x) is always real then

$$(Ax,x) = \overline{(Ax,x)} = \overline{(x,A^*x)} = (A^*x,x),$$

so that $([A-A^*]x,x) = 0$ for all x, and, by the theorem of the preceding section, $A = A^*$.

This theorem, as well as the theorem used in proving it, is false in real spaces. (Example?) For, in the first place, its proof depends on a theorem that is valid only in complex unitary spaces, and, in the second place, in a real space the reality of (Ax,x) (in fact of (Ax,y)) is a condition automatically satisfied by all A, whereas the condition $A = A^*$, or, equivalently, $(Ax,y) = (x,Ay)$, need not be satisfied.

Another proof of the thoroughgoing nature of our analogy is this fact: an arbitrary linear transformation A may be expressed, in one and only one way, in the form $A = B + iC$, where B and C are Hermitian. (We shall refer to B and C as the real and imaginary parts of A; the representation $A = B + iC$ is called the <u>Cartesian decomposition</u> of A). For if we write

(1) $$B = (1/2)(A + A^*),$$
$$C = (1/2i)(A - A^*),$$

then we have $B^* = (1/2)(A^* + A) = B$ and $C^* = (-1/2i)(A^* - A) = C$, and, of course, $A = B + iC$. From this proof of the existence of a Cartesian decomposition its uniqueness is also clear: if we do have $A = B + iC$, then $A^* = B - iC$ and consequently $A, B,$ and C are again connected by (1) and (2).

§55. ALGEBRAIC COMBINATIONS

It is quite easy to characterize the matrix of a Hermitian transformation A with respect to an orthogonal basis $\mathfrak{X} = (x_1, \ldots, x_n)$. If the matrix of A is (α_{ij}) then we know that the matrix of A^*, with respect to the dual basis of $\mathfrak{X}$, is (α_{ij}^*), where $\alpha_{ij}^* = \overline{\alpha_{ji}}$; but an orthogonal basis is self dual so that we have, (since $A = A^*$),

(1) $$\alpha_{ij} = \overline{\alpha_{ji}}.$$

We leave it to the reader to verify the converse: if (α_{ij}) is a matrix satisfying (1) then we may define the linear transformation A, by means of this matrix and an arbitrary orthogonal coordinate system $\mathfrak{X} = (x_1, \ldots, x_n)$, by the usual equations

$$A(\textstyle\sum_j \xi_j x_j) = \sum_i \eta_i x_i ,$$
$$\eta_i = \textstyle\sum_j \alpha_{ij} \xi_j ;$$

the condition (1) implies that the A so obtained is

Hermitian.

As a valuable exercise in the use of the inner product in $\mathfrak{P}$ that we defined in §44 the reader may wish to verify that the multiplication operator T, defined in (6) §20, is Hermitian whereas the differentiation operator D, defined in (4) §20, is not.

The algebraic rules for the manipulation of Hermitian transformations are easy to remember if we think of such transformations as the analogs of real numbers. Thus: if A and B are Hermitian, so is $A + B$; if A is Hermitian and $\alpha \neq 0$ then αA is Hermitian if and only if α is real; and if A has an inverse then A and A^{-1} are both or neither Hermitian. The place where something always goes wrong is multiplication: the product of two Hermitian transformations need not be Hermitian. However:

THEOREM 1. If A and B are Hermitian then AB and BA are Hermitian if and only if they are equal, (i.e. if and only if A and B commute).

PROOF. If $AB = BA$ then $(AB)^* = B^*A^* = BA = AB$. If $(AB)^* = AB$ then $(AB) = (AB)^* = B^*A^* = BA$.

THEOREM 2. If A is Hermitian then for an arbitrary B, B^*AB is Hermitian; if B has an inverse and B^*AB is Hermitian then so is A.

PROOF. If $A = A^*$, then $(B^*AB)^* = B^*A^*B^{**} = B^*AB$. If B has an inverse and B^*AB is Hermitian, then every vector x may be written in the form $x = By$, and since

$$(Ax,x) = (ABy,By) = (B^*ABy,y),$$

the reality of the last term for all y implies the

reality of the first term for all x.

§56. NON NEGATIVE TRANSFORMATIONS

When is a complex number ζ non negative? Two equally natural necessary and sufficient conditions are that ζ may be written in the form $\zeta = \tau^2$ with some real τ, or that ζ may be written in the form $\zeta = \sigma\bar{\sigma}$ with any σ. Remembering also the fact that the Hermitian character of a transformation A can be described in terms of the function (Ax,x), we may consider any one of the three conditions below and attempt to use it as the definition of a transformation being non negative.

(1) $A = B^2$, with some Hermitian B.

(2) $A = C^*C$, with some C.

(3) $(Ax,x) \geq 0$ for all x.

Before deciding which one of these three conditions to use as definition we observe that (1) $\Longrightarrow$ (2) $\Longrightarrow$ (3). For if $A = B^2$, $B = B^*$, then $A = BB = B^*B$, and if $A = C^*C$ then $(Ax,x) = (C^*Cx,x) = (Cx,Cx) = \|Cx\|^2 \geq 0$. It is actually true that (3) implies (1), so that the three conditions are equivalent, but we shall not be able to prove this till later. We adopt as our definition the third condition.

DEFINITION. A linear transformation A in <u>non negative</u>, in symbols $A \geq 0$, if for all x, $(Ax,x) \geq 0$.

More generally we shall write $A \geq B$ (or $B \leq A$) if $A-B \geq 0$. Although, of course, it is quite possible that the difference of two non Hermitian transformations is non negative, we shall generally use this notation for Hermitian transformations only.

Non negative transformations are usually called

non negative semi definite; if $A \geqq 0$ and $(Ax,x) = 0$ implies $x = 0$, A is called positive definite. Since the Schwarz inequality implies $|(Ax,x)| \leqq \|Ax\| \cdot \|x\|$, we see that for a positive definite operator A, $Ax = 0$ implies $x = 0$, (so that on a finite dimensional space a positive definite operator has an inverse). We shall see later that the converse is true: if $A \geqq 0$ and A has an inverse then A is positive definite. For positive definite transformations A we shall write $A > 0$; if $A - B > 0$ we also write $A > B$ (or $B < A$).

We observe that it follows from the theorem of §54 that if A, on a complex unitary space, is non negative, then A is Hermitian.

It is possible to give a matricial characterization of non negative transformations; we shall postpone this discussion until later. In the meantime we shall have occasion to refer to non negative matrices, meaning thereby matrices (α_{ij}) with the property that for every set $\{\xi_1, \ldots, \xi_n\}$ of n complex numbers we have $\sum_i \sum_j \alpha_{ij} \xi_i \bar{\xi}_j \geqq 0$. This condition is clearly equivalent to the condition that (α_{ij}) be the matrix, with respect to any orthogonal coordinate system, of a non-negative transformation.

The algebraic rules for non negative transformations are similar to those for Hermitian transformations as far as sums, scalar multiples, and inverses are concerned; even Theorem 2 (§55) remains valid if we replace "Hermitian" by "non-negative" throughout. It is also true that if A and B are non negative then AB and BA are non negative if and only if they are equal, so that A and B commute, but we shall have to postpone the proof of this statement till later.

§57. PERPENDICULAR PROJECTIONS

We are now in a position to fulfill our earlier promise to investigate the projections associated with the particular direct sum decompositions $\mathfrak{V} = \mathfrak{M} \oplus \mathfrak{M}^{\perp}$. We shall call such a projection a <u>perpendicular projection.</u> Since $\mathfrak{M}^{\perp}$ is uniquely determined by the linear manifold $\mathfrak{M}$, we need not specify both the direct summands associated with a projection if we already know that it is perpendicular. We shall call the (perpendicular) projection E on $\mathfrak{M}$ along $\mathfrak{M}^{\perp}$ simply the projection on $\mathfrak{M}$, and we shall write $E = P_{\mathfrak{M}}$

THEOREM 1. A linear transformation E is a perpendicular projection if and only if $E = E^2 = E^*$. Perpendicular projections are non negative linear transformations and have the property that $\| Ex \| \leqq \| x \|$ for all x.

PROOF. If E is a perpendicular projection then Theorem 1 (§33) and the theorem of §19 show (after, of course, the usual replacements, $\mathfrak{M}^{0} \longrightarrow \mathfrak{M}^{\perp}$, $A' \longrightarrow A^*$, etc.) that $E = E^*$. Conversely if $E = E^2 = E^*$, then the idempotence of E assures us that E is the projection on $\mathfrak{R}$ along $\mathfrak{N}$, where, of course, $\mathfrak{R} = \mathfrak{R}(E)$ and $\mathfrak{N} = \mathfrak{N}(E)$ are the range and null space of E respectively. Hence we need only show that $\mathfrak{R}$ and $\mathfrak{N}$ are orthogonal. For this purpose let x be any element of $\mathfrak{R}$ and y any element of $\mathfrak{N}$; the desired result follows from the relation

$$(x,y) = (Ex,y) = (x,E^*y) = (x,Ey) = 0.$$

The non negative nature of an E satisfying $E = E^2 = E^*$ follows from

$$(Ex,x) = (E^2x,x) = (Ex,E^*x) = (Ex,Ex) = \| Ex \|^2 \geqq 0.$$

Applying this result to the perpendicular projection $1 - E$ we see that

$$\| x \|^2 - \| Ex \|^2 = (x,x) - (Ex,x) = ([1-E]x,x) \geqq 0;$$

this concludes the proof of the theorem.

For some of the generalizations of our theory it is useful to know that idempotence together with the last property mentioned in Theorem 1 is also characteristic of perpendicular projections. In other words $E = E^2$ and $\| Ex \| \leqq \| x \|$ for all x imply $E = E^*$.

PROOF. We are to show that $\mathfrak{R}$ and $\mathfrak{N}$ are orthogonal. If x is in $\mathfrak{N}^\perp$ then $y = Ex-x$ is in $\mathfrak{N}$ since $Ey = E^2x - Ex = Ex - Ex = 0$. Hence $Ex = x + y$ with $(x,y) = 0$, so that $\| x \|^2 \geqq \| Ex \|^2 = \| x \|^2 + \| y \|^2 \geqq \| x \|^2$, and therefore $y = 0$. Consequently $Ex = x$ so that x is in $\mathfrak{R}$; $\mathfrak{N}^\perp \subset \mathfrak{R}$. Conversely if z is in $\mathfrak{R}$, $Ez = z$, we write $z = x + y$ with x in $\mathfrak{N}^\perp$ and y in $\mathfrak{N}$. Then $z = Ez = Ex + Ey = Ex = x$. ($Ex = x$ since x is in $\mathfrak{N}^\perp \subset \mathfrak{R}$). Hence z is in $\mathfrak{N}^\perp$, $\mathfrak{R} \subset \mathfrak{N}^\perp$, and therefore $\mathfrak{R} = \mathfrak{N}^\perp$.

We shall need also the fact that the theorem of §30 remains true if the word "projection" is qualified throughout by "perpendicular." This is an immediate consequence of the preceding characterization of perpendicular projections and of the fact that sums and differences of Hermitian transformations are Hermitian, whereas the product of two Hermitian transformations is Hermitian if and only if they commute. By the methods of unitary geometry it is also quite easy to generalize the part of this theorem dealing with sums from two to any finite number of summands. This generalization is most conveniently stated in terms of the concept of orthogonality for projections: we shall say that two (perpendicular) projections E and F are <u>orthogonal</u> if $EF = 0$. (Consideration of the adjoints shows that this is equivalent to $FE = 0$). That the geometric language is justified is shown by the following theorem.

THEOREM 2. Two perpendicular projections $E = P_{\mathfrak{M}}$ and $F = P_{\mathfrak{N}}$ are orthogonal if and only if the linear manifolds $\mathfrak{M}$ and $\mathfrak{N}$ (i.e.

the ranges of E and F) are orthogonal.

PROOF. If $EF = 0$, and if x and y are in the ranges of E and F respectively, then

$$(x,y) = (Ex,Fy) = (x,E^*Fy) = (x,EFy) = 0.$$

If conversely $\mathfrak{M}$ and $\mathfrak{N}$ are orthogonal, (so that $\mathfrak{N} \subset \mathfrak{M}^\perp$) then the fact that $Ex = 0$ for x in $\mathfrak{M}^\perp$ implies that $EFx = 0$ for all x (since Fx is in $\mathfrak{N}$ and consequently in $\mathfrak{M}^\perp$).

§58. ALGEBRAIC COMBINATIONS OF PERPENDICULAR PROJECTIONS

The sum theorem for perpendicular projections is now easy.

THEOREM 1. If $E_1, \ldots, E_n$ is a finite set of (perpendicular) projections, then $E = E_1 + \cdots + E_n$ is a (perpendicular) projection if and only if $E_iE_j = 0$ for every $i \neq j$, (i.e. if and only if the E_i are pairwise orthogonal).

PROOF. The proof that pairwise orthogonality implies that E is a projection is trivial; we prove explicitly only the converse so that we now assume that E is a perpendicular projection. Then for any x belonging to the range of E_i, for some fixed $i = 1, \ldots, n$, we have

$$\| x \|^2 \geq \| Ex \|^2 = (Ex,x) = (\textstyle\sum_j E_jx,x) = \sum_j (E_jx,x)$$
$$= \textstyle\sum_j \| E_jx \|^2 = \| E_ix \|^2 = \| x \|^2,$$

so that we must have equality all along. Since in particular we must have

$$\textstyle\sum_j \| E_jx \|^2 = \| E_ix \|^2,$$

we see that for $j \neq i$, $E_jx = 0$. In other words: any x in the range of E_i is in the null space (and consequently orthogonal to the range) of every E_j, for $j \neq i$; using Theorem 2 (§57) we draw the desired conclusion.

The straightforward generalization of (i), Theorem 1, (§33), (i.e. the statement obtained from Theorem 1 of the present section by omitting the parenthetical clauses) is also true, and is most easily proved by considering the traces of the summands and the sum; we do not enter into this proof here.

We conclude our discussion of projections with a remark on order relations. It is tempting to write $E \leqq F$ for two perpendicular projections $E = P_{\mathfrak{M}}$ and $F = P_{\mathfrak{N}}$, if $\mathfrak{M} \subset \mathfrak{N}$. Previously, however, we interpreted the sign $\leqq$ when used in an expression, such as $E \leqq F$, involving linear transformations to mean that $F - E$ is a non negative transformation. There are also other possible reasons for considering E to be smaller than F; we might have $\|Ex\| \leqq \|Fx\|$ for all x, or $FE = EF = E$, (see (ii), §30). The situation is straightened out by the following theorem, which plays here a role similar to Theorem 2 (§57), i.e. establishes the coincidence of several seemingly different concepts concerning projections, some of which are defined operatorially while others refer to the underlying geometrical objects.

THEOREM 2. For perpendicular projections $E = P_{\mathfrak{M}}$ and $F = P_{\mathfrak{N}}$ the following four conditions are equivalent.

(i) $E \leqq F$.

(ii) $\|Ex\| \leqq \|Fx\|$ <u>for all</u> x.

(iii) $\mathfrak{M} \subset \mathfrak{N}$

(iva) $FE = E$.

(ivb) $EF = E$.

PROOF. We shall prove the implication relations (i) $\Longrightarrow$ (ii) $\Longrightarrow$ (iii) $\Longrightarrow$ (iva) $\Longleftrightarrow$ (ivb) $\Longrightarrow$ (i).

(i) $\Longrightarrow$ (ii). If $E \leq F$ then for all x

$0 \leq ([F-E]x,x) = (Fx,x) - (Ex,x) = \| Fx \|^2 - \| Ex \|^2$, (since E and F are perpendicular projections).

(ii) $\Longrightarrow$ (iii). We assume $\| Ex \| \leq \| Fx \|$ for all x. Let us now take any x in $\mathfrak{M}$; then we have

$$\| x \| \geq \| Fx \| \geq \| Ex \| = \| x \|,$$

so that $\| Fx \| = \| x \|$ or $(x,x) - (Fx,x) = 0$, whence

$$([1 - F]x,x) = \| (1 - F)x \|^2 = 0,$$

and consequently $x = Fx$. In other words x in $\mathfrak{M}$ implies that x is in $\mathfrak{N}$, as was to be proved.

(iii) $\Longrightarrow$ (iva). If $\mathfrak{M} \subset \mathfrak{N}$, then for all x, Ex is in $\mathfrak{N}$, so that, for all x, $FEx = Ex$, as was to be proved.

That (iva) implies, and is in fact equivalent to, (ivb), follows by taking adjoints.

(iv) $\Longrightarrow$ (i). If $EF = FE = E$ then for all x

$(Fx,x) - (Ex,x) = (Fx,x) - (FEx,x) = (F[1-E]x,x)$.

Since E and F are commutative projections, so also are $(1-E)$ and F, and consequently $G = F(1-E)$ is a projection. Hence

$$(Fx,x) - (Ex,x) = (Gx,x) = \| Gx \|^2 \geq 0.$$

This completes the proof of Theorem 2.

In terms of the concepts introduced by now it is possible to give a quite intuitive sounding formulation of the Theorem of §30. Thus: for two perpendicular projections E and F, their sum, product, or difference is also a perpendicular projection if and only if F is respectively orthogonal to, commutative with, or greater than E.

§59. UNITARY TRANSFORMATIONS

We continue with our program of investigating tho analogy between numbers and transformations. When does a complex number ζ have absolute value one? Clearly a necessary and sufficient condition is that $\bar{\zeta} = 1/\zeta$; guided by our heuristic principle we are led to consider linear transformations U for which $U^* = U^{-1}$, or, equivalently, for which $UU^* = U^*U = 1$. Such transformations are called unitary. We observe that on a finite dimensional vector space either of the two conditions $UU^* = 1$ and $U^*U = 1$ implies the other, (see Theorems 1 and 2, §24). Concerning unitary transformations we prove the following theorem.

THEOREM. The following three conditions on a linear transformation U on a finite dimensional unitary space are equivalent to each other.

(1) U is unitary.

(2) $\| Ux \| = \| x \|$ for all x.

(3) $(Ux, Uy) = (x, y)$ for all x and y.

PROOF. If U is unitary then for all x

$$\| Ux \|^2 = (Ux, Ux) = (U^*Ux, x) = (x, x) = \| x \|^2.$$

If $\| Ux \| \equiv \| x \|$, we may use the identity

$$(4)\quad (x,y) = (1/4)\{\| x+y \|^2 - \| x-y \|^2 + i\| x+iy \|^2 - i\| x-iy \|^2\}.$$

Since the right side is invariant when we replace x and y by Ux and Uy, so is the left. (The identity (4) plays here a role similar to that of the polarization identity (1), §53: it enables us to pass from properties of $\| x \|^2$ to properties of (x, y).)

If, finally, $(Ux, Uy) \equiv (x, y)$ then

$$0 = (Ux, Uy) - (x, y) = (U^*Ux, y) - (x, y) = ([U^*U-1]x, y),$$

so that $U^*U = 1$. The finite dimensionality now assures us that UU^* is also 1, so that U is unitary. Since we have proved the implication relations (1) $\Longrightarrow$ (2) $\Longrightarrow$ (3) $\Longrightarrow$ (1), the proof of the theorem is complete; it is important to observe that (1) $\Longrightarrow$ (2) $\Longrightarrow$ (3) is true even in non finite dimensional spaces. We note also that U^{-1}

and U^* are unitary if and only if U is.

In any algebraic system, and in particular in general vector spaces and unitary spaces, it is of interest to consider the automorphisms of the system: i.e. to consider those one to one mappings of the system on itself which preserve all relations among its elements. We have seen already that the automorphisms of a general vector space are the non singular linear transformations. In a unitary space we require more of an automorphism, namely that it also preserve inner products (and consequently lengths). The preceding theorem shows that this requirement is equivalent to the condition that the transformation be unitary. Thus the two questions: "What linear transformations are the analogs of complex numbers of absolute value one?" and "What are the most general automorphisms of a unitary space?" have the same answer: unitary transformations. In the following section we shall show that unitary transformations furnish also the answer to a third important question.

§60. CHANGE OF ORTHOGONAL BASIS

We have seen that the theory of the passage from one linear basis of a vector space to another is best studied by means of an associated linear transformation A, (§§34,35); the question arises as to what special properties A has when we pass from one <u>orthogonal</u> basis of a unitary space to another. The answer is easy:

THEOREM 1. If $\mathfrak{X} = (x_1 . ,,, x_n)$ is an orthogonal basis of the n-dimensional unitary space $\mathfrak{V}$, and if U is any unitary transformation on $\mathfrak{V}$, then $U\mathfrak{X} = (Ux_1, ,,, Ux_n)$ is also an orthogonal basis of $\mathfrak{V}$. Conversely if U is a linear transformation and $\mathfrak{X}$ an orthogonal basis with the property that $U\mathfrak{X}$ is also an orthogonal basis then U is unitary.

PROOF. Since $(Ux_i, Ux_j) = (x_i, x_j) = \delta_{ij}$, $U\mathfrak{X}$ is an orthonormal set along with $\mathfrak{X}$; it is complete if $\mathfrak{X}$ is, since $(x, Ux_i) = 0$, for $i = 1, \ldots, n$, implies $(U^*x, x_i) = 0$, whence $U^*x = x = 0$. If, conversely, $U\mathfrak{X}$ is a complete orthonormal set along with $\mathfrak{X}$, then we have $(Ux, Uy) = (x, y)$ for all x and y in $\mathfrak{X}$, and it is clear that by linearity we obtain $(Ux, Uy) = (x, y)$ for all x, y.

We observe that the matrix (u_{ij}) of a unitary transformation U, with respect to an arbitrary orthogonal basis, satisfies the conditions

$$\sum_k u_{ik}\overline{u_{jk}} = \delta_{ij};$$

and, conversely, any such matrix together with any orthogonal basis, defines a unitary transformation. (Proof ?). As an other exercise in the use of unitary transformations the reader might prove that a linear transformation which satisfies any two of the three conditions of being involutory, Hermitian, or unitary, also satisfies the third, and that consequently the involutions associated ((A), §31) with perpendicular projections are also unitary.

An interesting and easy consequence of our considerations concerning unitary transformations is the following corollary of Theorem 1, §41.

THEOREM 2. Given any linear transformation A on an n-dimensional unitary space $\mathfrak{V}$, there exists an orthogonal basis $\mathfrak{X}$ in $\mathfrak{V}$ such that the matrix $[A; \mathfrak{X}]$ has the super diagonal form; or, equivalently, given any matrix $[A]$, there exists a unitary matrix $[U]$ such that $[U^{-1}][A][U]$ is superdiagonal.

PROOF. In the derivation of Theorem 2 from Theorem 1 (in §41) we constructed a (linear) basis $\mathfrak{X}$

$= (x_1, \ldots, x_n)$ with the property that $x_1, \ldots, x_j$ lies in $\mathfrak{M}_j$ and spans $\mathfrak{M}_j$ for $j = 1, \ldots, n$, and showed that in this basis the matrix of A is superdiagonal. If we knew that this basis is also an orthogonal basis we could apply Theorem 1 of the present section to obtain the desired result. But it is easy to make $\mathfrak{X}$ into an orthogonal basis even if it isn't one already: this is precisely what the Gram-Schmidt orthogonalization process (§48) can do. Here we use a special property of the Gram-Schmidt process, namely that the j-th element of the orthogonal basis it constructs is a linear combination of $x_1, \ldots, x_j$ and lies therefore in $\mathfrak{M}_j$.

We observe that in most of the preceding sections of this chapter we have treated complex unitary spaces. All of our theorems, past and future, have interesting and important analogs in the real case. In real vector spaces unitary transformations are called orthogonal and Hermitian ones symmetric. The main difference between the two disciplines is caused by the algebraic closure of the complex field. The closest counterpart of algebraic closure in the real case is the theorem that every polynomial can be factored into at most second degree pieces. Although we shall not treat the problems arising from this difference, we hope that by the time he has reached the end of this book the interested reader will be in a position where he will be able to apply the methods and results of our work to this more delicate study.

§61. CAYLEY TRANSFORM

The classes of complex numbers, whose transformation analogs we have been studying have certain algebraic relations to each other. One such relation is given by the complex valued function $\sigma = \frac{\tau - i}{\tau + i}$ of the real variable τ. This function maps the entire real line, $-\infty < \tau < \infty$, in a one to one way, on the unit circle minus the point $\sigma = 1$. (The geometric picture is quite easy: $\tau = 0$ becomes $\sigma = -1$, and the line is just wrapped around once to cover the circle.) The inverse mapping is given by $\tau = \frac{1}{i}\frac{\sigma + 1}{\sigma - 1}$. We shall show that the best possible analog of this result is valid in transformation theory.

THEOREM. If A is any Hermitian transformation on a finite dimensional unitary space $\mathfrak{B}$, then both the transformations $A \pm i1$ ($i = \sqrt{-1}$) have inverses; the transformation U defined by

(1) $U = (A+i1)^{-1}(A-i1) = (A-i1)(A+i1)^{-1}$

(called the <u>Cayley transform</u> of A) is unitary and fails to have the number 1 for a proper value. Consequently $U-1$ has an inverse and

(2) $A = \frac{1}{i}(U-1)^{-1}(U+1) = \frac{1}{i}(U+1)(U-1)^{-1}$.

Conversely, if U is any unitary transformation which fails to have the number 1 for a proper value then (2) defines a Hermitian transformation A and the Cayley transform of A is U.

PROOF. For any transformation A we have

$$((A \pm i1)x,(A \pm i1)y) = (Ax,Ay) \pm (Ax,iy) \pm (ix,Ay) + (x,y).$$

Hence if A is Hermitian we obtain, by taking $x = y$,
$\| (A + i1)x \|^2 = \|(A-i1)x\|^2 = \|Ax\|^2 + \|x\|^2$.
In other words $(A \pm i1)x = 0$ implies $x = 0$, so that both $A \pm i1$ have inverses and the definition (1) makes sense.

For any vector x we write $y = (A + i1)^{-1}x$, so that $x = (A + i1)y$; then

$\| Ux \| = \| (A-i1)y \| = \| (A + i1)y \| = \| x \|$;

U is unitary. Since, moreover, $Ux = x$ implies $(A - i1)y = (A + i1)y$, so that $y = x = 0$, 1 is not a proper value of U and, therefore, U-1 does have an inverse. Finally we have

$$x - Ux = (A + i1)y - (A - i1)y = 2iy,$$
$$x + Ux = (A + i1)y - (A - i1)y = 2Ay,$$

so that $A(x - Ux) = 2iAy = i(x + Ux)$; this establishes the validity of (2).

Let us now go backwards. Starting with U we define a transformation A by (2). For any pair of vectors x and y we write $x' = (U-1)^{-1}x, y' = (U-1)^{-1}y$, so that $x = (U-1)x', y = (U-1)y'$. Then $Ax = \frac{1}{i}(U+1)x'$, so that

$$\begin{aligned}(Ax,y) &= \frac{1}{i}\,(Ux' + x',\ Uy' - y') \\ &= \frac{1}{i}\,\{(Ux',Uy') + (x',Uy') - (Ux',y') - (x',y')\} \\ &= \frac{1}{i}\,\{(x',Uy') - (Ux',y')\} = i(Ux',y') + \overline{i(Uy',x')}.\end{aligned}$$

Since interchanging x and y replaces the last term of this relation by its own complex conjugate, it has the same effect on the first term, and consequently

$$(Ax,y) = \overline{(Ay,x)} = (x,Ay),$$

so that A is Hermitian.

If for any x we write $y = (U-1)^{-1}x$, so that $x = (U-1)y$, then $Ax = \frac{1}{i}(U+1)y$ and therefore

$$(A + i1)x = \frac{1}{i}\ \{(U+1)y - (U-1)y\} = \frac{1}{i}2y,$$

$$(A - i1)x = \frac{1}{i}\ \{(U+1)y + (U-1)y\} = \frac{1}{i}2Uy.$$

Consequently

$$U(A + i1)x = \frac{1}{i}\ 2Uy = (A - i1)x;$$

this establishes the validity of (1) and concludes the proof of the theorem.

It is worth remarking that even this theorem, as intimately as it may appear to be tied up with $i = \sqrt{-1}$, has a very natural analog in the real case. For if we write $B = iA$ then $U = (B-1)^{-1}(B+1)$ and, clearly, $B^* = -B$. Linear transformations with this latter property are called skew-Hermitian (or, in the real case, skew-symmetric). The natural and valid real analog of the theorem of this section establishes a correspondence, similar in every detail to the one we described, between orthogonal and skew-symmetric transformations.

§62. PROPER VALUES OF HERMITIAN AND UNITARY TRANSFORMATIONS

The analogy between numbers and transformations is supported even more than before by the following results which assert that the properties which caused us to define the special classes of transformations that we have been considering are reflected by their spectra.

THEOREM 1. If A is Hermitian then every proper value of A is real, and moreover, A is non negative or positive definite if and only if all its proper values are non negative or positive respectively.

PROOF. If $Ax = \lambda x$, with $x \neq 0$, then since A is Hermitian (Ax,x) is real and consequently

$$\frac{(Ax,x)}{\| x \|^2} = \lambda \frac{(x,x)}{\| x \|^2} = \lambda$$

is also real. The same proof establishes the statement concerning non negative transformations; the result for positive definite transformations follows from the fact that such a transformation must have an inverse and cannot therefore have the proper value zero.

THEOREM 2. Every proper value of a unitary transformation has absolute value one.

PROOF. If $Ux = \lambda x$, $x \neq 0$, then $\| x \| = \| Ux \| = | \lambda | \cdot \| x \|$.

Theorem 2.5 In spaces of odd dimension, Unitary transformation[s] have at least one real eigenvalue.

THEOREM 3. If A is either Hermitian or unitary then proper vectors belonging to different proper values are orthogonal.

PROOF. Suppose $Ax_1 = \lambda_1 x_1$, $Ax_2 = \lambda_2 x_2$, $\lambda_1 \neq \lambda_2$. Then if A is Hermitian we have

(1) $\lambda_1(x_1,x_2) = (Ax_1,x_2) = (x_1,Ax_2) = \lambda_2(x_1,x_2)$.

(The middle step makes use of the Hermitian character of A and the last step of the reality of λ_2). In case A is unitary (1) is replaced by

(2) $(x_1,x_2) = (Ax_1,Ax_2) = (\lambda_1/\lambda_2)(x_1,x_2)$,

(using the fact that $\bar{\lambda}_2 = 1/\lambda_2$). In either case $(x_1,x_2) \neq 0$ implies $\lambda_1 = \lambda_2$, so that we must have $(x_1,x_2) = 0$.

THEOREM 4. If a linear manifold $\mathfrak{M}$ reduces the unitary transformation U, on a finite dimensional unitary space, then so does $\mathfrak{M}^\perp$.

PROOF. Considered on the finite dimensional linear manifold $\mathfrak{M}$, U is a linear transformation with the property $(Ux,Uy) = (x,y)$; hence it is a unitary transformation on $\mathfrak{M}$ and as such has an inverse. Consequently every x in $\mathfrak{M}$ may be written in the form $x = Uy$ with y in $\mathfrak{M}$; in other words x in $\mathfrak{M}$ implies that $y = U^{-1}x$ is in $\mathfrak{M}$. Hence $\mathfrak{M}$ reduces $U^{-1} = U^*$. It follows from Theorem 2, §33, that $\mathfrak{M}^\perp$ reduces $(U^*)^* = U$.

We observe that the same result for Hermitian transformations (even in not necessarily finite dimensional spaces) is trivial, since if $\mathfrak{M}$ reduces A then $\mathfrak{M}^\perp$ reduces $A^* = A$.

THEOREM 5. If A is either a Hermitian

or a unitary transformation on an n-dimensional unitary space $\mathfrak{V}$, then the algebraic multiplicity of any proper value λ_0 of A is equal to its geometric multiplicity, i.e. to the dimension, say m, of the linear manifold $\mathfrak{M}$ of all solutions of $Ax = \lambda_0 x$.

PROOF. We shall use only the property described in Theorem 4 so that we may simultaneously establish the result for both the Hermitian and the unitary case.

It is clear that $\mathfrak{M}$, and therefore $\mathfrak{M}^{\perp}$, reduces A; let us denote by A_1 and A_2 the linear transformation A considered only on $\mathfrak{M}$ and $\mathfrak{M}^{\perp}$ respectively. By choosing a basis $(x_1, \ldots, x_n)$ in $\mathfrak{V}$ so that $x_1, \ldots, x_m$ are in $\mathfrak{M}$ and $x_{m+1}, \ldots, x_n$ are in $\mathfrak{M}^{\perp}$, we see that for all λ

$$\Delta(A - \lambda 1) = \Delta(A_1 - \lambda 1) \cdot \Delta(A_2 - \lambda 1).$$

Since A_1 is a linear transformation on an m-dimensional space with only one proper value λ_0, λ_0 must occur as a proper value of A_1 with the algebraic multiplicity m, so that $\Delta(A_1 - \lambda 1) = (\lambda_0 - \lambda)^m$. Since on the other hand λ_0 is not a proper value of A_2, so that $\Delta(A_2 - \lambda_0 1) \neq 0$, we see that $\Delta(A - \lambda 1)$ contains $(\lambda_0 - \lambda)$ as a factor exactly m times, as was to be proved.

§63. SPECTRAL THEOREM FOR HERMITIAN TRANSFORMATIONS

We are now ready to prove the main theorem of this book, the theorem of which most of the other results of this chapter are immediate corollaries. To a large extent what we have been doing up to the present was a matter of sport (useful, however, for generalizations): we wanted to show how much can conveniently be done with spectral theory before proving the spectral theorem. The spectral theorem, incidentally, can be made to follow trivially from the superdiagonalization process we have already described: because of the importance of the theorem we prefer to give below its (quite easy) direct proof. The reader may find it profitable to adapt the method of proof (not the result) of Theorem 2, §41, to prove as much as he can of the spectral theorem.

THEOREM. To any Hermitian linear transformation A on an n-dimensional unitary space there corresponds an integer p, $1 \leq p \leq n$, p perpendicular projections $E_1, \ldots, E_p$ (different from zero) and p numbers $\alpha_1, \ldots, \alpha_p$ with the following properties:

(1) the E_j are pairwise orthogonal,

(2) the α_j are pairwise different,

(3) $\sum_j E_j = 1$,

(4) $AE_j = E_j A$ for $j = 1, \ldots, p$,

(5) $\sum_j \alpha_j E_j = A$.

The α's and E's are uniquely determined by the conditions (1) - (5). The representation (5) is the spectral form of (A); some of its further important properties are:

(6) the α_j are exactly the distinct proper values of A and are consequently real;

(7) the dimension of the range of E_j is the multiplicity of α_j;

(8) a linear transformation B commutes with A if and only if it commutes with each E_j.

PROOF. Let $\alpha_1, \ldots, \alpha_p$ be the different proper values of A, and let $E_j, j = 1, \ldots, p$, be the perpendicular projection on the linear manifold of all solutions of $Ax = \alpha_j x$. Thus (2) and the first part of (6) are satisfied by definition; the second part of (6) follows from Theorem 1, §62, and (7) from Theorem 5, §62.

From Theorem 3, §62, we obtain (1) and also (3). (For (1) guarantees that $\sum_j E_j = E$ is a perpendicular projection; if it were not $= 1$ then A considered on the range of $1-E$ would be a linear transformation with no proper values). The truth of (4) follows from the fact that each $\mathfrak{M}_j$ reduces A; it remains to prove (5), (8), and uniqueness.

For any vector x we write $x_j = E_j x$; then x_j is in $\mathfrak{M}_j$ (so that $Ex_j = x_j$) and consequently $Ax_j = \alpha_j x_j$, $j = 1, \ldots, p$. It follows that

$$Ax = A(\sum_j E_j x) = \sum_j Ax_j = \sum \alpha_j x_j = \sum \alpha_j Ex_j = \sum \alpha_j E_j x;$$

this is precisely the statement of (5).

One half of (8) is clear from (5): if B commutes with each E_j it also commutes with A. Suppose on the other hand that B commutes with A, and consider any fixed E_{j_0}. B surely commutes with all polynomials in A, so that we will achieve our purpose if we can show that E_{j_0} is a polynomial in A. To do this, let us use (5) to see what a polynomial in A will look like. We have

$$A^2 = (\sum_i \alpha_i E_i)(\sum_j \alpha_j E_j) = \sum_i \sum_j \alpha_i \alpha_j E_i E_j = \sum_j \alpha_j^2 E_j$$

(since $E_i E_j = 0$ if $i \neq j$ and $E_j^2 = E_j$); similarly $A^n = \sum_j \alpha_j^n E_j$ for every positive integer n, and hence for any polynomial $p(\tau)$,

$$p(A) = \sum_j p(\alpha_j) E_j.$$

Now we are done: all we need to do is to find a polynomial $p(\tau)$ which is such that $p(\alpha_j) = 0$ for all $j \neq j_0$, and $p(\alpha_{j_0}) = 1$; for such a $p(\tau)$, $p(A) = E_{j_0}$. We may for example choose

$$p(\tau) = \prod_{j \neq j_0} \frac{\tau - \alpha_j}{\alpha_{j_0} - \alpha_j} .$$

To prove finally that the representation (5) is unique, we shall assume (1) - (5) and show first that the α_j are necessarily what we defined them to be in the existence proof, namely the distinct proper values of A. If x is any vector in the range of any E_j, so that $E_j x = x$ and $E_i x = 0$ for $i \neq j$, then

$$Ax = \sum_i \alpha_i E_i x = \alpha_j E_j x = \alpha_j x,$$

so that each α_j is a proper value of A. If conversely λ is any proper value of A, say $Ax = \lambda x$, $x \neq 0$, then we write $x_j = E_j x$ and we see that

$$Ax = \lambda x = \lambda \sum_j x_j \; .$$

and

$$Ax = A \sum_j x_j = \sum_j \alpha_j x_j$$

so that $\sum_j (\lambda - \alpha_j) x_j = 0$. Since the x_j are pairwise orthogonal, those among them that are not zero form a linearly independent set. It follows that for each j either $x_j = 0$ or else $\lambda = \alpha_j$. Since $x \neq 0$, we have $x_j \neq 0$ for some j, and consequently λ is indeed one of the α's. The rest of the uniqueness proof follows from the proof of (8): there we showed, using only (1) and (5), that each E_j is a polynomial in A, and that this polynomial is determined by the α's. This completes our proof of the spectral theorem.

Before exploiting this theorem we remark on its matricial interpretation. If we choose an orthogonal basis in the range of each E_j, then the totality of the vectors in these little bases forms a basis for the whole space: expressed in this basis the matrix (α_{ij}) of A will be <u>diagonal</u>, i.e. $\alpha_{ij} = \delta_{ij} \alpha_i$. The fact that by suitable choice of an orthogonal basis the matrix of a Hermitian transformation can be made diagonal, or, equivalently, that any Hermitian matrix [A] can be unitarily transformed (i.e. replaced by $[U]^{-1}[A][U]$) into a diagonal matrix, already follows of course from the superdiagonal form. We gave our operatorial version for two reasons. First it is this version which generalizes easily to the infinite dimensional case and, second, because we believe that even in the finite dimensional case, writing $\sum_i \alpha_i E_i$ often has great notational and typographical advantages over the matrix notation.

We shall also make use of the fact that a not necessarily Hermitian linear transformation A is unitarily diagonable (i.e. that its matrix with respect to a suitable orthogonal basis is diagonal) if and only if conditions (1) - (5) of the theorem of the present section hold for it. For if we have (1) - (5) then the proof of diagonability, given for Hermitian transformations, applies; the converse we leave as an exercise for the reader.

§64. NORMAL TRANSFORMATIONS

We have seen that every Hermitian transformation is diagonable and that an arbitrary transformation A may be written as $A = B + iC$ with B and C Hermitian; why isn't it true that simply by diagonalizing B and C separately we can diagonalize A? The answer is, of course, that diagonalization implies the choice of a suitable orthogonal basis and there is no reason to expect that a basis which diagonalizes B will have the same effect on C. It is of considerable importance to know the precise class of transformations for which the theorem of the preceding section is valid, and fortunately this class is easy to describe.

We shall call a linear transformation A <u>normal</u> if it commutes with its adjoint, $AA^* = A^*A$. We point out first that A is normal if and only if its real and imaginary parts commute. For suppose that A is normal and $A = B + iC$ with B and C Hermitian; since $B = (1/2)(A+A^*)$ and $C = (1/2i)(A-A^*)$ it is clear that $BC = CB$. If conversely $BC = CB$ then the relations $A = B + iC, A^* = B - iC$ imply that A is normal. We observe that Hermitian and unitary transformations are normal.

The class of transformations satisfying (1) - (5) of §63 is precisely the class of normal transformations. A half of this statement is easy to prove: if $A = \sum_j \alpha_j E_j$ then $A^* = \sum_j \bar{\alpha}_j E_j$, and it takes merely a simple computation to show that $AA^* = A^*A = \sum_j |\alpha_j|^2 E_j$. To prove the converse, i.e. that normality implies the existence of a spectral form we have two alternatives. We could derive this result from the spectral theorem for Hermitian transformations, using the real and imaginary parts of A, or we could prove that the essential lemmas of §62, on which the proof of the Hermitian case rests, are just as valid for an arbitrary normal operator. Because its methods are of some interest we adopt the second procedure. We observe that the machinery to prove the lemmas that follow was available to us in §62, so that we could have stated the spectral theorem for normal operators immediately; we travelled the present course in order to motivate the definition of normality.

THEOREM 1. If A is normal, then x is a proper vector of A if and only if it is a proper vector of A^*; if $Ax = \lambda x$ then $A^*x = \bar{\lambda}x$.

PROOF. We observe that the normality of A implies

(1) $$\| Ax \|^2 = (Ax, Ax) = (A^*Ax, x) = (AA^*x, x) = (A^*x, A^*x) = \| A^*x \|^2.$$

Since $A - \lambda 1$ is normal along with A, and since $(A - \lambda 1)^* = A^* - \bar{\lambda}1$, we obtain the relation

(2) $$\| Ax - \lambda x \| = \| A^*x - \bar{\lambda} x \|,$$

from which the assertions of the theorem follow immediately.

THEOREM 2. If A is normal then proper vectors belonging to different proper values are orthogonal.

PROOF. If $Ax_1 = \lambda_1 x_1, Ax_2 = \lambda_2 x_2$, then

$$\lambda_1(x_1,x_2) = (Ax_1,x_2) = (x_1,A^*x_2) = \lambda_2(x_1,x_2).$$

This theorem generalizes Theorem 3 of §62; in the proof of the spectral theorem we needed also Theorems 4 and 5 of §62. The following result takes the place of the first of these.

THEOREM 3. If A is normal, λ is a proper value of A, and $\mathfrak{M}$ is the set of all solutions of $Ax = \lambda x$, then both $\mathfrak{M}$ and $\mathfrak{M}^\perp$ reduce A.

PROOF. That $\mathfrak{M}$ reduces A we have seen before. To prove that $\mathfrak{M}^\perp$ also reduces A it is sufficient to prove that $\mathfrak{M}$ reduces A^*. This is easy: if x is in $\mathfrak{M}$ then

$$A(A^*x) = A^*(Ax) = \lambda(A^*x),$$

so that A^*x is also in $\mathfrak{M}$.

This theorem seems to be much weaker than its correspondent in §62. It is true that in a finite dimensional unitary space $A\mathfrak{M} \subset \mathfrak{M}$ implies $A\mathfrak{M}^\perp \subset \mathfrak{M}^\perp$ for a normal A, and it is even true that this property is characteristic of normality. (We shall not prove this: the reader may verify that it is a consequence of the spectral theorem for normal operators, which we shall prove). The most important thing to observe, however, is that the proof of Theorem 5 of §62 depended only on this weak property: the only manifolds that need be considered are the ones of the type mentioned in the preceding theorem.

This concludes the spade work: the spectral theorem

for normal operators follows just as before in the Hermitian case. If in the statement of the theorem of §63 we replace the word "Hermitian" by "normal" and delete the reference (in (6)) to the reality of the proper values, the rest of the statement and all of the proof remain unchanged.

It is the theory of normal operators that is of chief interest in the study of unitary spaces. Concerning normal operators it is useful to observe that spectral conditions of the type given in Theorems 1 and 2 of §62, there shown to be necessary for the Hermitian, unitary, etc. character of a transformation, are for normal operators also sufficient. Thus:

THEOREM 4. A normal transformation A with spectral form $A = \sum_j \alpha_j E_j$ is (1) Hermitian, (2) non negative, (3) positive definite, (4) unitary, (5) non singular, (6) idempotent if and only if all the α_j are (1') real, (2') non negative, (3') positive (4') of absolute value one, (5') not zero, (6') zero or one.

PROOF. Since we know that the α_j are the proper values of A we know also (j) implies (j') for $j = 1, \ldots, 6$. Since $A^* = \sum_j \bar{\alpha}_j E_j$, we see that (1') implies (1). If $\alpha_j \geqq 0$ then for any x we have

$$(Ax, x) = \sum_j \alpha_j (E_j x, x) = \sum_j \alpha_j \| E_j x \|^2 \geqq 0,$$

so that (2') implies (2) and also (3') implies (3). (Proof: $\sum_j \alpha_j \| E_j x \|^2 = 0$ implies that for each j either $\alpha_j = 0$ or $E_j x = 0$, and since $\alpha_j \neq 0$, $E_j x = 0$ for $j = 1, \ldots, p$, so that $x = \sum_j E_j x = 0$). To prove that (4') implies (4) we observe that (4') implies

$$AA^* = A^*A = \sum_j | \alpha_j |^2 E_j = \sum_j E_j = 1.$$

If $\alpha_j \neq 0$ for $j = 1, \ldots, p$, we may form the linear transformation $B = \sum_j 1/\alpha_j E_j$; it is clear that $AB = BA = 1$, so that A is non singular. Finally $A^2 = \sum_j \alpha_j^2 E_j$ so that if $\alpha_j^2 = \alpha_j$ for $j = 1, \ldots, p$, then $A^2 = A$.

We observe that the implication relations (5') $\Rightarrow$ (5), (2) $\Rightarrow$ (2'), and (3') $\Rightarrow$ (3) together prove an assertion we made in §61; if A is non negative and non singular then it is positive definite.

§65. FUNCTIONS OF NORMAL TRANSFORMATIONS

One of the most useful concepts for normal operators is that of a function of an operator. If A is a normal linear transformation with spectral form $A = \sum_j \alpha_j E_j$ and if $f(\zeta)$ is an arbitrary complex valued function of the complex variable ζ, defined at least for $\zeta = \alpha_j$, $j = 1, \ldots, p$, then we define a linear transformation $f(A)$ by

$$f(A) = \sum_j f(\alpha_j) E_j.$$

Since for polynomials (and even for rational functions) $p(\zeta)$ we have already seen that our earlier definition of $p(A)$ yields, for a normal A, $p(A) = \sum_j p(\alpha_j) E_j$, we see that the new notion is a generalization of the old one. The advantage of considering $f(A)$ for arbitrary functions f is for us largely notational: it introduces nothing conceptually new. For we may write for an arbitrary $f(\zeta)$, $f(\alpha_j) = \beta_j$, and then we may find a polynomial $p(\zeta)$ which at the finite set of distinct complex numbers α_j takes, respectively, the values β_j. With this polynomial $p(\zeta)$ we have $f(A) = p(A)$, so that the class of transformations defined by $f(A)$ is nothing essentially new: it only saves the trouble of constructing a polynomial $p(\zeta)$ to fit each special case. Thus for example if for every complex number λ we define $f_\lambda(\zeta) = 1$ if $\zeta = \lambda$, $f_\lambda(\zeta) = 0$ otherwise, then $f_\lambda(A)$ = the perpendicular projection on the linear manifold of solutions of $Ax = \lambda x$.

We observe that if $f(\zeta) = 1/\zeta$ then (assuming of course that $f(\zeta)$ is defined for all α_j, i.e. that $\alpha_j \neq 0$) $f(A) = A^{-1}$, and if $f(\zeta) = \bar{\zeta}$ then $f(A) = A^*$. These statements imply that if $f(\zeta)$ is an arbitrary rational function of ζ and $\bar{\zeta}$, we obtain $f(A)$ by the replacements $\zeta \longrightarrow A$, $\bar{\zeta} \longrightarrow A^*$, $1/\zeta \longrightarrow A^{-1}$, $\alpha \longrightarrow \alpha 1$. The symbol $f(A)$ is, however, defined for much more general functions and we shall in what follows make free use of expressions such as e^A and $\sqrt{A}$.

As an exercise in the use of the functional calculus the reader may wish to prove the theorem of §61 (concerning the Cayley transform of a Hermitian transformation) by considering the function $f(\zeta) = \frac{\zeta - i}{\zeta + i}$. Consideration of the function $f(\zeta) = e^{i\zeta}$ shows similarly that for every Hermitian A, $U = e^{iA}$ is unitary, and that conversely every unitary U has the form $U = e^{iA}$ with a Hermitian A.

§66. PROPERTIES OF NON NEGATIVE TRANSFORMATIONS

A particularly important function is the square root of non negative operators. We consider $f(\zeta) = \sqrt{\zeta}$, defined for all real $\zeta \geqq 0$ as the non negative square root of ζ, and for every non negative $A = \sum_j \alpha_j E_j$, $\alpha_j \geqq 0$, we consider

$$f(A) = \sqrt{A} = \sum_j \sqrt{\alpha_j} E_j.$$

It is clear that $\sqrt{A} \geqq 0$ and that $(\sqrt{A})^2 = A$; we should like to investigate the extent to which these properties characterize $\sqrt{A}$. At first glance it may seem hopeless to look for any uniqueness since if we consider $B = \sum_j \pm \sqrt{\alpha_j} E_j$, with an arbitrary choice of sign in each place, we still have $A = B^2$. The $\sqrt{A}$ we constructed, however, was non negative, and wo can show that this additional property guarantees uniqueness: in other words $A = B^2$, $B \geqq 0$, implies $B = \sqrt{A}$. For let

$B = \sum_k \beta_k F_k$ be the spectral form of B; then

$$\sum_k \beta_k^2 F_k = B^2 = A = \sum_j \alpha_j E_j.$$

Since the β_k are distinct and non negative so also are the β_k^2; the uniqueness of the spectral form of A implies that each β_k^2 is equal to some α_j (and conversely), and that the corresponding E's and F's are equal. By a permutation of the indices we may therefore achieve $\beta_j^2 = \alpha_j$, $j = 1, \ldots, p$, so that $\beta_j = \sqrt{\alpha_j}$, as was to be shown.

There are several important applications of the existence of square roots for non negative operators of which we now give two.

First: we recall that in §56 we mentioned three possible definitions of a non negative transformation and adopted the weakest one, namely that $(Ax, x) \geqq 0$ for all x. The strongest of the three possible definitions was that we could write A in the form $A = B^2$ with a Hermitian B; we point out that the result of this section concerning square roots implies that the (seemingly) weakest of our conditions implies and is therefore equivalent to the strongest. (In fact we can even achieve a unique non negative B!)

Second: in §56 we stated also that if A and B are non negative and commutative then AB is also non-negative; we can now give an easy proof of this assertion. The commutativity of A and B implies that any two of the transformations A, B, $\sqrt{A}$, $\sqrt{B}$ commute with each other; consequently

$$AB = \sqrt{A}\ \sqrt{A}\ \sqrt{B}\ \sqrt{B} = \sqrt{A}\ \sqrt{B}\ \sqrt{A}\ \sqrt{B} = (\sqrt{A}\ \sqrt{B})^2.$$

Since $\sqrt{A}$ and $\sqrt{B}$ are Hermitian and commutative, their product is Hermitian and therefore its square is non negative.

The spectral theory also makes it quite easy to characterize the matrix (with respect to an arbitrary orthogonal coordinate system) of a non negative transformation A. Since the determinant $\Delta(A)$ is the product of the proper values of A it is clear that $A \geq 0$ implies $\Delta(A) \geq 0$. If we consider the defining property of non negativeness expressed in terms of the matrix (α_{ij}) of A, i.e. $\sum_{ij} \alpha_{ij} \xi_i \bar{\xi}_j \geq 0$, we observe that this last expression remains non negative if we restrict the coordinates $\{\xi_1, \ldots, \xi_n\}$ by requiring that a certain number of them vanish. In terms of the matrix this means that if we cross out columns numbered $j_1, \ldots, j_k$, say, and cross out also the rows bearing the <u>same</u> numbers, the remaining small matrix is still non negative, and consequently so is its determinant. This fact is usually expressed by saying that the <u>principal minors</u> of the determinant of a non negative matrix are non negative. The converse is true; the coefficient of the j-th power of λ in the characteristic polynomial $\Delta(A-\lambda 1)$ of A is (except for sign) the sum of all principal minors of n-j rows and columns. The sign is alternately plus and minus; this implies that if A has non negative principal minors and is Hermitian (so that the zeros of $\Delta(A-\lambda 1)$ are known to be real) then the proper values of A are non negative. (Proof ?) Since the Hermitian character of a matrix is ascertainable by observing whether or not the elements α_{ij} are Hermitian symmetric $(\alpha_{ij} = \bar{\alpha}_{ji})$, our comments reduce the problem of finding out whether or not a matrix is non negative to a finite number of elementary computations.

Using the above characterization of non negativeness the reader may verify that if $A = \begin{bmatrix} 1 & 0 \\ 0 & 0 \end{bmatrix}$ and $B = \begin{bmatrix} 0 & 0 \\ 0 & 1 \end{bmatrix}$ and if C is a Hermitian matrix for which both $A \leq C$ and $B \leq C$ then

$$C = \begin{bmatrix} 1+\epsilon & \sqrt{\epsilon(1+\epsilon)}\,\theta \\ \sqrt{\epsilon(1+\epsilon)}\,\bar{\theta} & 1+\epsilon \end{bmatrix} \leftarrow \text{error read} \quad \begin{bmatrix} 1+\epsilon & \theta \\ \bar{\theta} & 1+\delta \end{bmatrix}$$

where $\epsilon \geq 0$ and ~~$|\theta| = 1$~~. It is also easy to show

$\delta \geq 0$ and $|\theta|^2 \leq \min\{\epsilon(1+\delta), \delta(1+\epsilon)\}$

that for ~~two matrices C_1 and C_2~~ a matrix of the type of C, $C \leqq 1$ ~~$C_1 \geqq C_2$~~ can ~~never~~ hold ~~unless $C_1 = C_2$~~ if and only if $C = 1$. In modern terminology these facts together show that Hermitian matrices with the ordering induced by the notion of non negativeness do not form a <u>lattice</u>. Restricting attention to the real case and interpreting a matrix $\begin{bmatrix} \alpha & \beta \\ \beta & \gamma \end{bmatrix}$ as the point $\{\alpha, \beta, \gamma\}$ in three dimensional space, the ordering and its non lattice character take on an amusing geometric aspect.

§67. POLAR DECOMPOSITION

There is another useful consequence of the theory of square roots, namely the analog of the polar representation $\zeta = \rho e^{i\theta}$ of a complex number.

> THEOREM 1. If A is an arbitrary linear transformation on a finite dimensional unitary space $\mathfrak{V}$, then there is a (uniquely determined) non negative transformation P, and a unitary transformation U, such that $A = UP$. U is uniquely determined by A if and only if A is non singular.

PROOF. Although it is not logically necessary to do so we shall first give the proof in the case where A has an inverse: the proof in the other case is an obvious modification of this proof, which gives greater insight into the geometric structure of arbitrary transformations.

Since the transformation A^*A is non negative we may find its (unique) non negative square root, $P = \sqrt{A^*A}$. We write $V = PA^{-1}$; since $VA = P$ the theorem will be proved if we can prove that V is unitary, for then we may write $U = V^{-1}$. Since $V^* = (A^{-1})^* P^* = (A^*)^{-1} P$, we see that

$$V^*V = (A^*)^{-1}\ PPA^{-1} = (A^*)^{-1}\ A^*AA^{-1} = 1,$$

so that (since $\mathfrak{V}$ is finite dimensional) V is unitary, and we are done. To prove uniqueness we observe that $UP = U_oP_o$ implies $PU^* = P_oU_o^*$ and therefore

$$P^2 = PU^*UP = P_oU_o^*U_oP_o = P_o^2 .$$

Since the non negative transformation $P^2 = P_o^2$ has only one non negative square root it follows that $P = P_o$. (In this proof we did not use the fact that A has an inverse). If A is non singular then so is P (since $P = U^{-1}A$), and from this we obtain (multiplying the relation $UP = U_oP_o$ on the right by $P^{-1} = P_o^{-1}$) $U = U_o$.

We turn now to the general case, where we do not assume that A^{-1} exists. We form P exactly the same way as in the preceding proof, so that $P^2 = A^*A$, and then we observe that for any vector x we have

$$\| Px \|^2 = (Px,Px) = (P^2x,x) = (A^*Ax,x) = \| Ax \|^2.$$

If for each vector y in the range $\mathfrak{R}(P)$ of $P, y = Px$, we define $Uy = Ax$, the transformation U is length preserving wherever it is defined. We must show that U is uniquely determined: i.e. that $Px_1 = Px_2$ implies $Ax_1 = Ax_2$. This is true since $P(x_1-x_2) = 0$ is equivalent to $\| P(x_1-x_2) \| = 0$ and this latter condition implies $\| A(x_1-x_2) \| = 0$. If we define U on the orthogonal complement of $\mathfrak{R}(P)$ to be, say, the identity, then the transformation U, thereby determined on all $\mathfrak{V}$, is unitary and has the property that $UPx = Ax$ for all x. In other words $A = UP$, as was to be proved. Incidentally, the extent of non uniqueness of U is clear from the proof, in which at one place we were free to make an almost entirely arbitrary choice for U. (As long as U is unitary, its behavior on $(\mathfrak{R}(A))^{\perp}$ is immaterial).

Applying the theorem just proved to A^* in place of A, and then taking adjoints, we obtain also the dual fact that every A may be written in the form $A = PU$ with a unitary U and a non negative P.

In geometric language this theorem is sometimes stated in the following form: every linear transformation on $\mathfrak{V}$ is effected by a dilatation followed by a rotation. (The justification for the terminology is clear from the diagonal forms of the matrices of non negative and unitary transformations). The reader might give an alternative proof of this theorem, for the special case of normal transformations, by using the spectral form.

In contrast with the Cartesian decomposition (§54) we call the representation $A = UP$ the polar decomposition of A; in terms of this decomposition we obtain a new characterization of normality.

THEOREM 2. If $A = UP$ is the polar decomposition of the linear transformation A then a necessary and sufficient condition that A be normal is that $UP = PU$.

PROOF. (Since U is not necessarily uniquely determined by A this statement is to be interpreted as follows: if A is normal then P commutes with every U, and if P commutes with any single U then A is normal).

Since $AA^* = UP^2U^* = UP^2U^{-1}$ and $A^*A = P^2$, it is clear that A is normal if and only if U commutes with P^2. Since, however, P^2 is a function of P and conversely P is a function of P^2, ($P = \sqrt{P^2}$), it follows that commuting with P^2 is equivalent to commuting with P.

§68. PROBLEMS OF COMMUTATIVITY

The spectral theory of normal operators and the functional calculus may also be used to solve certain problems concerning commutativity. This is a deep and extensive subject: more to illustrate the method than for the actual results we discuss two theorems from it.

> THEOREM 1. Two Hermitian transformations A and B on a finite dimensional unitary space are commutative if and only if there exists a Hermitian transformation C and two real valued functions of a real variable, say f and g, such that $A = f(C)$, $B = g(C)$. If such a C exists then we may even choose C in the form $C = h(A,B)$, where h is a suitable real valued function of two real variables.

PROOF. The sufficiency of the condition is clear; we prove only the necessity.

Let $A = \sum_i \alpha_i E_i$ and $B = \sum_j \beta_j F_j$ be the spectral forms of A and B; since A and B commute it follows from (8), §63, that E_i and F_j commute. Let $h(s,t)$ be any function of the two real variables s and t for which the numbers $h(\alpha_i, \beta_j) = \gamma_{ij}$ are all distinct, and write $C = h(A,B) = \sum_i \sum_j h(\alpha_i, \beta_j) E_i F_j$. (It is clear that h may even be chosen as a polynomial, and the same will be true about the functions f and g we are about to describe). Let f and g be such that $f(\gamma_{ij}) = \alpha_i$ and $g(\gamma_{ij}) = \beta_j$ for all i and j. Then $f(C) = A$ and $g(C) = B$, and everything is proved.

> THEOREM 2. If A is a normal transformation on a finite dimensional unitary space and if B is an arbitrary transformation which commutes with A then B commutes with A^*.

PROOF. Let $A = \sum_i \alpha_i E_i$ be the spectral form of A; then $A^* = \sum_i \bar{\alpha}_i E_i$. Let $f(\zeta)$ be such a function (polynomial) of the complex variable ζ that $f(\alpha_i) = \overline{\alpha}_i$ for all i. Then $A^* = f(A)$ and the theorem follows.

Theorem 2 is remarkable in two ways. It asserts in the first place a kind of transitivity for commutativity. It is not true in general that if A_1 commutes with A and A with B then A_1 will commute with B; but for $A_1 = A^*$ this is precisely the statement of Theorem 2. The other remarkable feature of this theorem is its reluctance to be generalized: its truth or falsity has not yet been decided for a very large class of operators in infinite dimensional spaces.

§69. HERMITIAN TRANSFORMATIONS OF RANK ONE

We have already seen (Theorem 2, §38) that every linear transformation A of rank ρ is the sum of ρ linear transformations of rank one. It is easy to see (using the spectral theorem) that if A is Hermitian, or non negative, then the summands may also be taken Hermitian, or non negative, respectively. We know (Theorem 1, §38) what the matrix of a transformation of rank one has to be; what more can we say if the transformation is Hermitian or non negative ?

THEOREM 1. If A has rank one and is Hermitian (or non negative) then in every orthogonal coordinate system the matrix (α_{ij}) of A has the form $\alpha_{ij} = \kappa\beta_i\overline{\beta}_j$ with a real κ, (or the form $\alpha_{ij} = \gamma_i\overline{\gamma}_j$). If, conversely, [A] has this form in a single orthogonal coordinate system then A has rank one and is Hermitian (or non negative).

PROOF. We know that the matrix (α_{ij}) of a transformation A of rank one, in any orthogonal coordinate system $\mathfrak{X} = (x_1, \ldots, x_n)$, has the form $\alpha_{ij} = \beta_i \gamma_j$. If A is Hermitian we must also have $\alpha_{ij} = \overline{\alpha_{ji}}$ whence $\beta_i \gamma_j = \overline{\beta_j \gamma_i}$. If for some i, $\beta_i = 0$ and $\gamma_i \neq 0$ then for all j, $\overline{\beta_j} = \beta_i \gamma_j / \gamma_i = 0$, whence $A = 0$. Since we assumed that the rank of A is one (and not zero) this is impossible. Similarly $\beta_i \neq 0$ and $\gamma_i = 0$ is impossible; i.e. we can find an i for which $\beta_i \gamma_i \neq 0$. Using this i we have $\overline{\beta_j} = (\beta_i / \overline{\gamma_i})\ \gamma_j = \kappa \gamma_j$ with some constant κ independent of j. Since the diagonal elements $\alpha_{jj} = (Ax_j, x_j) = \beta_j \gamma_j$ of a Hermitian matrix are real, we can even conclude that in this case $\alpha_{ij} = \kappa \beta_i \overline{\beta_j}$ with a real κ.

If, moreover, A is non negative, then we even know that $\kappa \beta_j \overline{\beta_j} = \alpha_{jj} = (Ax_j, x_j)$ is non negative, and therefore so is κ. In this case we write $\lambda = \sqrt{\kappa}$, and the relation $\kappa \beta_i \overline{\beta_j} = (\lambda \beta_i)(\overline{\lambda \beta_j})$ shows that α_{ij} has the form $\alpha_{ij} = \gamma_i \overline{\gamma_j}$.

It is easy to see that these necessary conditions are also sufficient. If $\alpha_{ij} = \kappa \beta_i \overline{\beta_j}$ with a real κ then A is Hermitian. If $\alpha_{ij} = \gamma_i \overline{\gamma_j}$, and $x = \sum_i \xi_i x_i$ then

$$(Ax, x) = \sum_i \sum_j \alpha_{ij} \overline{\xi_i}\, \xi_j = \sum_i \sum_j \gamma_i \overline{\gamma_j}\, \overline{\xi_i}\, \xi_j$$

$$= (\sum_i \gamma_i \overline{\xi_i})(\overline{\sum_j \xi_j \overline{\xi_j}}) = |\sum_i \gamma_i \overline{\xi_i}|^2 \geq 0,$$

so that A is non negative.

As a consequence of Theorem 1 it is very easy to prove a remarkable theorem on non negative matrices.

THEOREM 2. If A and B are non negative linear transformations whose matrices in some orthogonal coordinate system are (α_{ij}) and (β_{ij}) respectively then the linear transformation C, whose matrix (γ_{ij}) in this same coordinate system is given by $\gamma_{ij} = \alpha_{ij} \beta_{ij}$ for all i and j, is also non negative. (The matrix (γ_{ij}) is called the <u>Hadamard product</u> of (α_{ij}) and (β_{ij}).)

PROOF. Since we may write both A and B as a sum of non negative transformations of rank one, so that $\alpha_{ij} = \sum_p \alpha_i^p \overline{\alpha}_j^p$ and $\beta_{ij} = \sum_q \beta_i^q \overline{\beta}_j^q$, we have

$$\gamma_{ij} = \sum_p \sum_q \alpha_i^p \beta_i^q \overline{(\alpha_j^p \beta_j^q)}.$$

Since a sum of non negative matrices is non negative, it will be sufficient to prove that, for each fixed p and q, $(\alpha_i^p \beta_i^q)\overline{(\alpha_j^p \beta_j^q)}$ defines a non negative matrix: and this follows from Theorem 1. This proof shows by the way that Theorem 2 remains valid if we replace "non negative" by "Hermitian" in both hypothesis and conclusion; for later applications, however, it is only the actually stated version that will be useful to us.

§70. CONVERGENCE OF VECTORS

Essentially the only way in which we exploited, so far, the existence of an inner product in unitary spaces was to introduce the notion of a normal transformation. A much more obvious circle of ideas is the study of the convergence problems that arise in a unitary space.

Let us see what we might mean by the assertion that a sequence $\{x_n\}$ of vectors in $\mathfrak{V}$ converges to a vector x in $\mathfrak{V}$. There are two possibilities that suggest themselves:

(i) $\| x_n - x \| \longrightarrow 0$ as $n \longrightarrow \infty$;

(ii) $(x_n - x, y) \longrightarrow 0$ as $n \longrightarrow \infty$, for each fixed y in $\mathfrak{V}$.

If (i) is true then we have for every y

$$|(x_n - x, y)| \leq \| x_n - x \| \cdot \| y \| \longrightarrow 0,$$

so that (ii) is true. In a finite dimensional space the converse implication is valid: (ii) $\Longrightarrow$ (i). To prove this let $z_1, \ldots, z_N$ be an orthogonal basis in $\mathfrak{V}$. (Often in the remainder of this chapter we shall write N for the dimension of a finite dimensional vector space, in order to reserve n for the dummy variable in limiting processes). If we assume (ii) then, for each $i = 1, \ldots, N$, $(x_n - x, z_i) \longrightarrow 0$. Since (Theorem 2, §47)

$$\| x_n - x \|^2 = \sum_i |(x_n - x, z_i)|^2$$

it follows that $\| x_n - x \| \longrightarrow 0$, as was to be proved.

Concerning the convergence of vectors (in either of the two equivalent senses) we shall use without proof the following facts. (All these facts are easy consequences of our definitions and the properties of convergence in the usual domain of complex numbers: we assume that the reader has a modicum of familiarity with these notions).

The expression $\alpha x + \beta y$ is a continuous function of all its arguments simultaneously; i.e. if $\{\alpha_n\}$ and $\{\beta_n\}$ are sequences of complex numbers and $\{x_n\}$ and $\{y_n\}$ are sequences of vectors, then $\alpha_n \longrightarrow \alpha$, $\beta_n \longrightarrow \beta$, $x_n \longrightarrow x, y_n \longrightarrow y$, implies that $\alpha_n x_n + \beta_n y_n \longrightarrow \alpha x + \beta y$. If $\{z_i\}$ is an orthogonal basis in $\mathfrak{V}$, $x_n = \sum_i \alpha_{in} z_i$, $x = \sum_i \alpha_i z_i$, then $x_n \longrightarrow x$ if and only if $\alpha_{in} \longrightarrow \alpha_i$ (as $n \longrightarrow \infty$) for each $i = 1, \ldots, N$. (Thus the notion of convergence here defined coincides with the usual one in N-dimensional complex Euclidean space). Finally we shall assume as known the fact that a finite dimensional unitary space $\mathfrak{V}$ with the metric $\| x - y \|$ is complete: i.e. if $\{x_n\}$ is a sequence of vectors for which $\| x_n - x_m \| \longrightarrow 0$ as $n, m \longrightarrow \infty$, then there is a unique vector x such that $x_n \longrightarrow x$ as $n \longrightarrow \infty$.

§71. BOUND OF A LINEAR TRANSFORMATION

The metric properties of vectors have certain important implications for the metric properties of linear transformations, which we now begin to study.

DEFINITION. A linear transformation A on a unitary space $\mathfrak{U}$ is <u>bounded</u> if there exists a positive finite constant K such that for every vector x in $\mathfrak{U}$, $\| Ax \| \leq K \| x \|$. The greatest lower bound of all constants K with this property is called the <u>bound</u> of A, in symbols $|\!|\!| A |\!|\!|$; it is clear that for every x, $\| Ax \| \leq |\!|\!| A |\!|\!| \cdot \| x \|$.

For examples we may consider the cases where A is a perpendicular projection $\neq 0$ or a unitary transformation: Theorem 1 of §57 and the theorem of §59 respectively imply that in both cases $|\!|\!| A |\!|\!| = 1$. Consideration of the vectors $x_n = x_n(t) \equiv t^n$ in $\mathfrak{P}$ shows that the differentiation operator is not bounded.

Because in the sequel we shall have occasion to consider quite a few upper and lower bounds similar to $|\!|\!| A |\!|\!|$ we introduce a convenient notation. If P is any possible property of real numbers t, we shall denote the set of all real numbers t possessing property P by the symbol $\{t : P\}$, and we shall denote greatest lower bound and least upper bound by <u>inf</u> (for infimum) and <u>sup</u> (for supremum) respectively. In this notation for example

$$|\!|\!| A |\!|\!| = \inf \{K : \| Ax \| \leq K \| x \| \text{ for all } x\}.$$

The notion of boundedness is closely connected with the notion of continuity. If A is bounded and if ϵ is any positive number, by choosing $\delta = \epsilon / |\!|\!| A |\!|\!|$ we may make sure that $\| x-y \| \leq \delta$ implies

$$\| Ax-Ay \| = \| A(x-y) \| \leq |\!|\!| A |\!|\!| \cdot \| x-y \| \leq \epsilon ;$$

in other words boundedness implies (uniform) continuity. It is true that the best possible converse of this result is valid: continuity of A at any single point implies that A is bounded and consequently uniformly continuous over the whole space. Since we shall not need this result we leave its proof to the interested reader: we turn rather to the proof that in the case of chief interest to us boundedness is always present.

THEOREM. Every linear transformation A on a finite dimensional unitary space $\mathfrak{B}$ is bounded.

PROOF. Let $(x_1, \ldots, x_N)$ be an orthogonal basis in $\mathfrak{B}$ and write

$$K_0 = \max \{\| Ax_1 \|, \ldots, \| Ax_N \|\}.$$

Since an arbitrary vector x may be written in the form $x = \sum_i (x,x_i)x_i$ we obtain, applying the Schwarz inequality and remembering that $\| x_i \| = 1$,

$$\begin{aligned} \| Ax \| = \| A(\textstyle\sum_i (x,x_i)x_i)\| &= \| \textstyle\sum_i (x,x_i)Ax_i \| \\ &\leqq \textstyle\sum_i |(x,x_i)| \cdot \| Ax_i \| \leqq \\ &\leqq \textstyle\sum_i \| x \| \cdot \| x_i \| \cdot \| Ax_i \| \\ &\leqq K_0 \textstyle\sum_i \| x \| = NK_0 \| x \|. \end{aligned}$$

In other words $K = NK_0$ is a bound of A.

It is no accident that the dimension N of $\mathfrak{B}$ enters into our evaluation: we have already seen that the theorem is not true in non finite dimensional spaces.

§72. EXPRESSIONS FOR THE BOUND

To facilitate working with the bound of a transformation we consider the following four expressions:

$$p = \sup\{\| Ax \| / \| x \| : x \neq 0\},$$

$$q = \sup\{\| Ax \| : \| x \| = 1\},$$

$$r = \sup\{|(Ax,y)| / \| x \| \cdot \| y \| : x \neq 0,\ y \neq 0\},$$

$$s = \sup\{|(Ax,y)| : \| x \| = \| y \| = 1\}.$$

In accordance with our definition of the brace notation the expression $\{\| Ax \| : \| x \| = 1\}$, for example, means the set of all real numbers of the form $\| Ax \|$, considered for all x's for which $\| x \| = 1$).

Since $\| Ax \| \leq K \| x \|$ is trivially true with any non negative K if $x = 0$, the definition of sup implies that $p = ||| A |||$; we shall prove that in fact $p = q = r = s = ||| A |||$. Since the supremum in the expression for q is being extended over a subset of the corresponding set for p, (i.e. if $\| x \| = 1$ then $\| Ax \| / \| x \| = \| Ax \|$), we see that $q \leq p$; a similar argument shows that $s \leq r$.

For any $x \neq 0$ we consider $y = x/\| x \|$, (so that $\| y \| = 1$); we have $\| Ax \|/\| x \| = \| Ay \|$. In other words every number of the set whose supremum is p, occurs also in the corresponding set for q; it follows that $p \leq q$, and consequently $p = q = ||| A |||$.

Similarly if $x \neq 0$ and $y \neq 0$ we consider $x' = x/\| x \|$ and $y' = y/\| y \|$; we have

$$|(Ax,y)|/\| x \| \cdot \| y \| = |(Ax',y')|,$$

and hence, by the argument just used, $r \leq s$, so that $r = s$.

To consolidate our position: we have proved so far that

$$p = q = ||| A |||, \; r = s.$$

Since

$$\frac{|(Ax,y)|}{\| x \| \cdot \| y \|} \leq \frac{\| Ax \| \cdot \| y \|}{\| x \| \cdot \| y \|} = \frac{\| Ax \|}{\| x \|}$$

it follows that $r \leq p$; we shall complete the proof by showing that $p \leq r$. For this purpose we consider any vector x for which $Ax \neq 0$ (so that also $x \neq 0$); for such an x we write $y = Ax$ and we have

$$\| Ax \| / \| x \| = |(Ax,y)|/ \| x \| \cdot \| y \|.$$

In other words we proved that every number which occurs in the set defining p, and which is not zero, occurs also in the set of which r is the supremum: this clearly implies the desired result.

The numerical function $||| A |||$ of operators A satisfies the following four relations:

(1) $$|||A + B||| \leq |||A||| + |||B|||,$$

(2) $$|||AB||| \leq |||A||| \cdot |||B|||,$$

(3) $$|||\alpha A||| = |\alpha| \cdot |||A|||,$$

(4) $$|||A^*||| = |||A|||.$$

The proof of the first three of these is immediate from the definition of a bound; for the proof of (4) we use the equation $|||A||| = r$, as follows. Since

$$|(Ax,y)| = |(x,A^*y)| \leq \|x\| \cdot \|A^*y\|$$
$$\leq |||A^*||| \cdot \|x\| \cdot \|y\|,$$

we see that $|||A||| \leq |||A^*|||$; replacing A by A^* and A^* by $A^{**} = A$ we obtain the reverse inequality.

§73. BOUNDS OF A HERMITIAN TRANSFORMATION

As usual we can say a little more about the special case of Hermitian transformations than in the general case. We consider for any Hermitian transformation A the sets of real numbers

$$\Phi = \{(Ax,x)/\|x\|^2 : x \neq 0\},$$
$$\Psi = \{(Ax,x) : \|x\| = 1\}.$$

It is clear that $\Psi \subset \Phi$. If for every $x \neq 0$ we write $y = x/\|x\|$ then $\|y\| = 1$ and $(Ax,x)/\|x\|^2 = (Ay,y)$, so that every number in Φ occurs also in Ψ, and consequently $\Phi = \Psi$. We write

$$\alpha = \inf \Phi = \inf \Psi,$$
$$\beta = \sup \Phi = \sup \Psi;$$

α is the <u>lower bound</u> and β the <u>upper bound</u> of the Hermitian transformation A. If we recall the definition of non negative transformations we see that α is the greatest real number for which $A - \alpha 1 \geq 0$, and β is the least real number for which $\beta 1 - A \geq 0$. Concerning these numbers we assert that

$$\gamma = \max\{|\alpha|, |\beta|\} = |||A|||.$$

PROOF. Since $|(Ax,x)| \leqq \| Ax \| \cdot \| x \| \leqq ||| A ||| \cdot \| x \|^2$, it is clear that $|\alpha|$ and $|\beta|$ are both $\leqq ||| A |||$. To prove the reverse inequality we observe that the non negative character of the two linear transformations $\gamma 1 - A$ and $\gamma 1 + A$ implies that both

$$(\gamma 1 + A)^*(\gamma 1 - A)(\gamma 1 + A) = (\gamma 1 + A)(\gamma 1 - A)(\gamma 1 + A)$$

and

$$(\gamma 1 - A)^*(\gamma 1 + A)(\gamma 1 - A) = (\gamma 1 - A)(\gamma 1 + A)(\gamma 1 - A)$$

are non negative, and therefore so also is their sum $2\gamma(\gamma^2 1 - A^2)$. Since $\gamma = 0$ implies $||| A ||| = 0$ the theorem is trivial in this case; in any other case we may divide by 2γ and obtain the result that $(\gamma^2 1 - A^2) \geqq 0$. In other words

$$\gamma^2 \| x \|^2 = \gamma^2 (x,x) \geqq (A^2x,x) = \| Ax \|^2,$$

whence $\gamma \geqq ||| A |||$, and the proof is complete.

We call the reader's attention to the fact that the computation in the main body of this proof could have been avoided entirely. Since $\gamma 1 - A$ and $\gamma 1 + A$ are both non negative, and since they commute, we may con clude immediately (§66) that their product $\gamma^2 1 - A^2$ is non negative. We presented this roundabout method in accordance with the principle that, with an eye to the generalizations of our theory, one should avoid using the spectral theory whenever possible. Our proof of the fact that the non negativeness and commutativity of A and B imply $AB \geqq 0$ was based on the existence of square roots for non negative transformations. This theorem can also be proved by so called "elementary" methods, i.e. methods not using the spectral theorem, but even the simplest elementary proof involves complications which are purely technical and for our purposes not particularly useful.

§74. MINIMAX PRINCIPLE

A very elegant and useful fact concerning Hermitian transformations is the following minimax principle, due to R. Courant.

THEOREM. Let A be a Hermitian transformation on an n-dimensional unitary space $\mathfrak{V}$, and let $\lambda_1, \ldots, \lambda_n$ be the (not necessarily distinct) proper values of A, with the notation so chosen that $\lambda_1 \geq \lambda_2 \geq \cdots \geq \lambda_n$. For any linear manifold $\mathfrak{M}$ in $\mathfrak{V}$ we write

$$\mu(\mathfrak{M}) = \sup \{(Ax,x) : x \text{ in } \mathfrak{M}, \|x\| = 1\}$$

and for $k = 1, \ldots, n$ we define

$$\mu_k = \inf \{\mu(\mathfrak{M}) : \text{dimension of } \mathfrak{M} = n-k+1\}.$$

Then $\mu_k = \lambda_k$ for $k = 1, \ldots, n$.

PROOF. Let $x_1, \ldots, x_n$ be an orthogonal basis in $\mathfrak{V}$ for which $Ax_i = \lambda_i x_i$, $i = 1, \ldots, n$ (§63); let $\mathfrak{M}_k$ be the linear manifold spanned by $x_1, \ldots, x_k$, $k = 1, \ldots, n$. Since the dimension of $\mathfrak{M}_k$ is k, $\mathfrak{M}_k$ cannot be disjoint from any $n - k + 1$ dimensional linear manifold $\mathfrak{M}$ in $\mathfrak{V}$; if $\mathfrak{M}$ is any such manifold we may find a vector x belonging to both $\mathfrak{M}_k$ and $\mathfrak{M}$ and such that $\|x\| = 1$. For this $x = \sum_{i=1}^{k} \xi_i x_i$ we have

$$(Ax,x) = \sum_{i=1}^{k} \lambda_i |\xi_i|^2 \geq \lambda_k \sum_{i=1}^{k} |\xi_i|^2 = \lambda_k \|x\|^2 = \lambda_k,$$

so that $\mu(\mathfrak{M}) \geq \lambda_k$.

If on the other hand we consider the particular $(n - k + 1)$-dimensional linear manifold $\mathfrak{M}_0$ spanned by $x_k, x_{k+1}, \ldots, x_n$, then for any $x = \sum_{i=k}^{n} \xi_i x_i$ in this manifold we have (assuming $\|x\| = 1$)

$$(Ax,x) = \sum_{i=k}^{n} \lambda_i |\xi_i|^2 \leq \lambda_k \sum_{i=k}^{n} |\xi_i|^2 = \lambda_k \|x\|^2 = \lambda_k,$$

so that $\mu((\mathfrak{M}_0) \leq \lambda_k$.

In other words, as $\mathfrak{M}$ runs over all $n - k + 1$ dimensional linear manifolds, $\mu(\mathfrak{M})$ is always $\geq \lambda_k$

and is at least once $\leqq \lambda_k$; this shows that $\mu_k = \lambda_k$, as was to be proved.

In particular for $k = 1$ we see (using §73) that for a Hermitian transformation A, $|||A|||$ is equal to the absolute value of the greatest proper value of A. That this is not true for all linear transformations is seen by considering the matrix $\begin{pmatrix} 0 & 1 \\ 0 & 0 \end{pmatrix}$. (See comment at the end of §35. This particular matrix has practically no properties and is very useful, for this reason, in the construction of counter examples).

§75. CONVERGENCE OF LINEAR TRANSFORMATIONS

We return now to the consideration of convergence problems. There are three obvious senses in which we may try to define the convergence of a sequence $\{A_n\}$ of linear transformations to a fixed linear transformation A.

(i) $|||A_n - A||| \longrightarrow 0$ as $n \to \infty$.

(ii) $\| A_n x - Ax \| \longrightarrow 0$ as $n \to \infty$ for each fixed x.

(iii) $|(A_n x, y) - (Ax, y)| \longrightarrow 0$ as $n \to \infty$ for each fixed x and y.

If (i) is true then for every x

$$\| A_n x - Ax \| = \| (A_n - A) x \| \leqq |||A_n - A||| \cdot \| x \| \longrightarrow 0,$$

so that (i) $\Longrightarrow$ (ii). We have already seen (§70) that (ii) $\Longrightarrow$ (iii) and that in finite dimensional spaces (iii) $\Longrightarrow$ (ii). It is even true that in finite dimensional spaces (ii) $\Longrightarrow$ (i), so that all three conditions become equivalent. To prove this, let $x_1, \ldots, x_N$ be an orthogonal basis in $\mathfrak{V}$; we suppose that (ii) is valid. Then for any $\epsilon > 0$ we may find an $n_0 = n_0(\epsilon)$ such that for $n \geqq n_0$, $\| A_n x_i - Ax_i \| \leqq \epsilon$ for $i = 1, \ldots, N$. It follows that for an arbitrary $x = \sum_i (x, x_i) x_i$ we have

$$\| (A_n - A)x \| = \| \sum_i (x, x_i)(A_n - A)x_i \|$$
$$\leqq \sum_i \| x \| \cdot \| (A_n - A)x_i \| \leqq \epsilon N \| x \|,$$

and this implies (i).

It is also easy to prove that, using $|||A - B|||$ as a distance for operators, the resulting metric space is complete, i.e. that if $|||A_n - A_m||| \longrightarrow 0$ as $n,m \longrightarrow \infty$ then there is an A such that $|||A_n - A||| \longrightarrow 0$. The proof of this fact is reduced to the corresponding statement for vectors. If $|||A_n - A_m||| \longrightarrow 0$, then for each x, $\| A_n x - A_m x \| \longrightarrow 0$, so that we may find a vector corresponding to x which we may denote by, say, Ax, such that $\| A_n x - Ax \| \longrightarrow 0$. It is clear that the correspondence from x to Ax is given by a linear transformation A; the implication relation (ii) $\Longrightarrow$ (iii), proved above, completes the proof.

Now that we know what convergence means for linear transformations it behooves us to examine some simple functions of these transformations in order to verify their continuity. We assert that $|||A|||$, $\|Ax\|$, (Ax,y), Ax, $A + B$, αA, AB, and A^* are all continuous functions of all their arguments simultaneously. (Observe that the first three are numerical valued functions, the next is vector valued, and the last four are operator valued). The proofs of these statements are all quite easy, and similar to each other; to illustrate the ideas we discuss $|||A|||$, Ax, and A^*.

(1) If $A_n \longrightarrow A$, i.e. $|||A_n - A||| \longrightarrow 0$ then since the relations

$$|||A_n||| \leq |||A_n - A||| + |||A||| ,$$

$$|||A||| \leq |||A - A_n||| + |||A_n||| ,$$

imply that

$$\Big| \, |||A_n||| - |||A||| \, \Big| \leq |||A_n - A||| ,$$

we see that $|||A_n||| \longrightarrow |||A|||$.

(2) If $A_n \longrightarrow A$ and $x_n \longrightarrow x$ then

$$\| A_n x_n - Ax \| \leq \| A_n x_n - Ax_n \| + \| Ax_n - Ax \| \longrightarrow 0$$

so that $A_n x_n \longrightarrow Ax$.

(3) If $A_n \longrightarrow A$ then for each x and y

$$(A^*_n x, y) = (x, A_n y) = \overline{(A_n y, x)} \longrightarrow \overline{(Ay, x)}$$
$$= \overline{(y, A^*x)} = (A^*x, y),$$

whence $A_n^* \longrightarrow A^*$.

§76. ERGODIC THEOREM FOR UNITARY TRANSFORMATIONS

The routine work being out of the way, we illustrate the general theory by considering some very special but quite important convergence problems. The first of these is the ergodic theorem for unitary transformations.

THEOREM. Let U be a unitary transformation on a finite dimensional unitary space $\mathfrak{V}$, and let $\mathfrak{M}$ be the linear manifold of all solutions of $Ux = x$. Then the sequence $V_n = (1/n)(1+U+\cdots+U^{n-1})$ converges as $n \longrightarrow \infty$ to the perpendicular projection $E = P_{\mathfrak{M}}$.

PROOF. Let $\mathfrak{N}$ be the range of the linear transformation $1 - U$. For any x in $\mathfrak{N}$, $x = y - Uy$, we have

$$V_n x = (1/n)(y - Uy + Uy - U^2y + \cdots + U^{n-1}y - U^ny)$$
$$= (1/n)(y - U^ny),$$

so that

$$\| V_n x \| = (1/n)\| y - U^ny \| \leq (1/n)(\| y \| + \| U^ny \|)$$
$$= \frac{2\| y \|}{n}.$$

Hence for x in $\mathfrak{N}$, $V_n x$ converges to zero.

On the other hand if x is in $\mathfrak{M}$, i.e. $Ux = x$, then $V_n x = x$, so that in this case $V_n x$ certainly converges to x.

We shall complete the proof by showing that $\mathfrak{N}^\perp = \mathfrak{M}$ (This will imply that every vector is a sum of two vectors for which V_n converges, so that V_n converges everywhere. What we have already proved about the limit of V_n in $\mathfrak{M}$ and $\mathfrak{N}$ shows that $V_n x$ will always converge to the projection of x into $\mathfrak{M}$). To show that $\mathfrak{N}^\perp = \mathfrak{M}$ we observe that x is in the orthogonal complement of $\mathfrak{N}$ if and only if $(x,y - Uy) = 0$ for all y. This in turn implies that

$$0 = (x,y - Uy) = (x,y) - (x,Uy) = (x,y) - (U^*x,y)$$
$$= (x - U^*x,y),$$

i.e. that $x - U^*x = x - U^{-1}x$ is orthogonal to <u>every</u> vector y, so that $x - U^{-1}x = 0, x = U^{-1}x$, or $Ux = x$. Reading the last computation from right to left shows that this necessary condition is also sufficient; we need only recall the definition of $\mathfrak{M}$ to see that $\mathfrak{M} = \mathfrak{N}^\perp$

This very ingenious proof, which works with only very slight modifications in most of the important infinite dimensional cases, is due to F. Riesz. As an amusing exercise, which will show how one might have been led to think of the Riesz proof, the reader may wish to give an alternative proof based on the spectral theorem for unitary operators.

§77. POWER SERIES

We consider next the so called Neumann series, $\sum_{n=0}^{\infty} A^n$, where A is a linear transformation of bound < 1 on a finite dimensional vector space. We write

$$(1) \qquad S_p = \sum_{n=0}^{p} A^n,$$

then

$$(2) \qquad (1 - A)S_p = S_p - AS_p = 1 - A^{p+1}.$$

To prove that S_p has a limit as $p \to \infty$ we consider (for any $p > q$)

$$\|S_p - S_q\| \leq \sum_{n=q+1}^{p} \|A^n\| \leq \sum_{n=q+1}^{p} \|A\|^n.$$

Since $\|A\| < 1$ the last written number approaches zero

as $p,q \longrightarrow \infty$; it follows that S_p has a limit S as $p \longrightarrow \infty$. To evaluate this limit we observe that $1 - A$ has an inverse, since $(1 - A)x = 0$ implies that $Ax = x$ and consequently implies (unless $x = 0$) the impossibility

$$\| Ax \| = \| x \| > ||| A ||| \cdot \| x \| .$$

Hence we may write (2) in the form

$$(3) \quad S_p = (1 - A^{p+1})(1 - A)^{-1} = (1 - A)^{-1}(1 - A^{p+1});$$

since $A^{p+1} \longrightarrow 0$ as $p \longrightarrow \infty$, it follows that $S = (1 - A)^{-1}$.

As another example of infinite series of operators we consider the exponential series. For an arbitrary linear transformation A (not necessarily with $||| A ||| < 1$) we write

$$S_p = \sum_{n=0}^{p} (1/n!) A^n.$$

Since we have

$$||| S_p - S^q ||| \leq \sum_{n=q+1}^{p} (1/n!) \, ||| A |||^n$$

and since the right member, being a part of the power series for $\exp(||| A |||) = e^{||| A |||}$, converges to zero as $p,q \longrightarrow \infty$, we see that there is a linear transformation S such that $S_p \longrightarrow S$. We write $S = \exp(A)$; we shall merely mention some of the elementary properties of this function of A.

Consideration of the superdiagonal forms of A and S_p, shows that the proper values of $\exp(A)$ are, including multiplicities, the exponentials of the proper values of A. From the consideration of the superdiagonal form it follows also that the determinant of $\exp(A)$, i.e. $\prod_{i=1}^{N} \exp(\lambda_i)$, where $\lambda_1, \ldots, \lambda_n$ are the (not necessarily distinct) proper values of A, is the same as $\exp(\lambda_1 + \cdots + \lambda_n) = \exp(T(A))$. Since $\exp(\zeta) \neq 0$, this shows incidentally that $\exp(A)$ is always non singular. Considered as a function of A, $\exp(A)$ retains many of the simple properties of the ordinary numerical exponential function. Let us, for example, take any two commutative linear transformations A and B. Since $\exp(A + B) - \exp(A)\exp(B)$ is the limit (as $p \longrightarrow \infty$) of the expression

$$\sum_{n=0}^{p}\left(\frac{1}{n!}\right)(A+B)^n - \sum_{m=0}^{p}\left(\frac{1}{m!}\right)A^m \cdot \sum_{k=0}^{p}\left(\frac{1}{k!}\right)B^k$$
$$= \sum_{n=0}^{p}\left(\frac{1}{n!}\right)\sum_{j=0}^{n}\binom{n}{j} A^j B^{n-j} - \sum_{m=0}^{p}\sum_{k=0}^{p}\left(\frac{1}{m!k!}\right)A^m B^k,$$

we will have proved the multiplication rule for exponentials when we have proved that this expression converges to zero. (Here $\binom{n}{j}$ stands for the combinatorial coefficient $\frac{n!}{j!(n-j)!}$.) An easy verification yields the fact that for $k + m \leqq p$, $A^m B^k$ occurs in both terms of the last written expression with coefficients which differ only in sign; the terms that do not cancel out are all in the subtrahend and are together equal to

$$\sum_m \sum_k \frac{1}{m!k!} A^m B^k,$$

the summation being extended over those values of m and k which are $\leqq p$ and for which $m + k > p$. Since $m + k > p$ implies that at least one of the two integers m and k is greater than the integer part of $p/2$ (in symbols $[p/2]$), the bound of this remainder is dominated by

$$\sum_{m=0}^{\infty}\sum_{k=[p/2]}^{\infty} \frac{1}{m!k!} ||| A |||^m ||| B |||^k$$
$$+ \sum_{k=0}^{\infty}\sum_{m=[p/2]}^{\infty} \frac{1}{m!k!} ||| A |||^m \cdot ||| B |||^k$$
$$= \left(\sum_{m=0}^{\infty} \frac{1}{m!} ||| A |||^m\right)\left(\sum_{k=[p/2]}^{\infty} \frac{1}{k!} ||| B |||^k\right)$$
$$+ \left(\sum_{k=0}^{\infty} \frac{1}{k!} ||| B |||^k\right)\left(\sum_{m=[p/2]}^{\infty} \frac{1}{m!} ||| A |||^m\right)$$
$$= \exp(||| A |||)\, \alpha_p + \exp(||| B |||)\, \beta_p,$$

where $\alpha_p \longrightarrow 0$ and $\beta_p \longrightarrow 0$ as $p \longrightarrow \infty$.

Similar methods serve to discuss $f(A)$ where $f(\zeta)$

is any function representable by a power series,

$$f(\zeta) = \sum_{n=0}^{\infty} \alpha_n \zeta^n,$$

and where $|\!|\!| A |\!|\!|$ is smaller than the radius of convergence of the series. We leave it to the reader to verify that the functional calculus we are here hinting at is consistent with the functional calculus for normal transformations. Thus for example exp(A) as defined above is the same linear transformation as is defined by our previous notion of exp(A) for normal A's.

APPENDIX I

THE CLASSICAL CANONICAL FORM

The chief difference between the first two chapters and the third chapter of this book is that in the third chapter we picked out, by means of properties of the inner product in a unitary space, certain special classes of transformations and concentrated our attention on them. These classes are all subclasses of the class of normal transformations, and, using the spectral theorem, we see that the chief virtue of these transformations is that their structure is completely known if we know what one dimensional linear manifolds they leave invariant (i.e. what their proper vectors are). In order to understand the structure of not necessarily normal transformations we must study higher dimensional invariant linear manifolds. It turns out that the results here are of the same degree of difficulty in a general vector space as in a unitary space; we return accordingly to the consideration of arbitrary linear transformations in arbitrary (finite dimensional) vector spaces. The one restriction that we retain is that the scalars should be elements of an algebraically closed field such as the field of complex numbers. We begin by discussing two seemingly irrelevant and rather special notions (quotient space and nilpotent transformation); these concepts are, however, useful for studying the structure of linear transformations as well as in many other parts of the theory.

If $\mathfrak{V}$ is any vector space and $\mathfrak{M}$ is any linear manifold in $\mathfrak{V}$ we say that two vectors x and y of $\mathfrak{V}$ are <u>congruent</u> modulo $\mathfrak{M}$, in symbols $x \equiv y$ ($\mathfrak{M}$), if $x - y$ is in $\mathfrak{M}$. (This notion of congruence is a very close analog of the corresponding notion in number theory, according to which two integers are called congruent modulo m if their difference is a multiple of m.) We observe that $x \equiv 0\ (\mathfrak{M})$ is equivalent to x being in $\mathfrak{M}$, and that $x_1 \equiv y_1\ (\mathfrak{M})$ and $x_2 \equiv y_2\ (\mathfrak{M})$ imply that $\alpha_1 x_1 + \alpha_2 x_2 \equiv \alpha_1 y_1 + \alpha_2 y_2\ (\mathfrak{M})$. For any x_0 in $\mathfrak{V}$ let us denote by x_0^* the set of all vectors x of $\mathfrak{V}$ for which $x \equiv x_0\ (\mathfrak{M})$. (Since we are not going to consider unitary spaces in this appendix we are at liberty to use the star in a different sense from the customary one). Let us denote by $\mathfrak{V}^* = \mathfrak{V}/\mathfrak{M}$ the class of sets x_0^* (called congruence or residue classes) so obtained; we shall introduce linear operations into $\mathfrak{V}^*$ in such a way that $\mathfrak{V}^*$ will become a vector space. We define $\alpha_1 x_1^* + \alpha_2 x_2^* = (\alpha_1 x_1 + \alpha_2 x_2)^*$; the only thing we must make sure about is that this definition uniquely determines $(\alpha_1 x_1 + \alpha_2 x_2)^*$. If, in other words, $y_1^* = x_1^*$, and $y_2^* = x_2^*$, then we must be sure that $(\alpha_1 y_1 + \alpha_2 y_2)^* = (\alpha_1 x_1 + \alpha_2 x_2)^*$. This is true: the reader may verify that it is implied by the linearity of the congruence relation. The space $\mathfrak{V}^*$ is called the <u>quotient space</u> of $\mathfrak{V}$ modulo $\mathfrak{M}$. This notion is very rich in interesting properties which we shall not explore; we hope that the techniques we have developed in this book will enable the reader to ask and answer the relevant questions, such for example as the relation of quotient spaces to conjugate spaces, annihilators, etc. We merely describe the little concerning linear transformations that we shall have occasion to use.

If A is a linear transformation on $\mathfrak{V}$, and if $\mathfrak{M}$ is a linear manifold invariant under A, we define a linear transformation A^* on $\mathfrak{V}^*$ by $A^*x^* = (Ax)^*$. It is again easy to verify that A^* is uniquely defined. What we are interested in is this: if A happens not only to be reduced but also to be completely reduced by

$\mathfrak{M}$, so that A becomes the direct sum $A_1 \oplus A_2$ of two linear transformations defined on the subspaces $\mathfrak{M}$ and $\mathfrak{N}$ of $\mathfrak{V}$ respectively, then what is the relation between A_2 and A^*? Both these transformations can be considered as complementary to A_1; A_1 describes what A does on $\mathfrak{M}$, and both A_2 and A^* describe in different ways what A does elsewhere.

Let x and y be any two elements of $\mathfrak{N}$, and consider the corresponding elements x^* and y^* of $\mathfrak{V}^* = \mathfrak{V}/\mathfrak{M}$. If it should happen that $x^* = y^*$, so that $(x - y)^* = 0$, then this means that $x - y$ is in $\mathfrak{M}$ and $\mathfrak{N}$ at the same time and consequently $x = y$. Since on the other hand every element x of $\mathfrak{V}$ has the form $x = y + z$, with y in $\mathfrak{M}$ and z in $\mathfrak{N}$, we see that $x^* = z^*$, so that every x^* in $\mathfrak{V}^*$ is the star of some element in $\mathfrak{N}$. In other words the correspondence $x \to Tx = x^*$ is a one to one correspondence between $\mathfrak{N}$ and $\mathfrak{V}^*$; it follows from the definition of x^* that T is a linear transformation. (Thus we obtain in particular that $\mathfrak{N}$ and $\mathfrak{V}^*$ are isomorphic vector spaces). Let us now compare the transformation A_2 (i.e. A on $\mathfrak{N}$) with A^*. If $A_2x = y$, then $A^*x^* = (Ax)^* = y^*$; in other words $A^*Tx = Ty = TA_2x$. This implies that $A^*T = TA_2$ or $A^* = TA_2T^{-1}$. Loosely speaking (see §35) we may say that A^* transforms $\mathfrak{V}^*$ the same way as A_2 transforms $\mathfrak{N}$. In particular all characteristic features (such as proper values, invariant manifolds, etc.) are shared by A_2 and A^*; as linear transformations they are abstractly identical (isomorphic). We shall exploit this fact presently; at the moment we turn to our next topic.

A linear transformation A is called nilpotent of index q if $A^q = 0$ and $A^{q-1} \neq 0$, for some positive integer q. Concerning nilpotent transformations we shall need the following not particularly exciting but quite useful theorem.

THEOREM 1. If A is nilpotent of index q, and x_0 is any vector for which $A^{q-1}x_0 \neq 0$, then the vectors $x_0, Ax_0, \ldots, A^{q-1}x_0$ are linearly independent. If we write $\mathfrak{H}$ for the linear manifold spanned by these vectors then there exists a linear manifold $\mathfrak{K}$ such that $\mathfrak{V} = \mathfrak{H} \oplus \mathfrak{K}$ and such that A is completely reduced by the pair $(\mathfrak{H}, \mathfrak{K})$.

PROOF. We prove first the asserted linear independence. If $\sum_{i=0}^{q-1} \alpha_i A^i x_0 = 0$, let j be the first integer for which $\alpha_j \neq 0$. (We do not exclude the possibility $j = 0$). Then dividing through by $-\alpha_j$ and changing notation in an obvious way we obtain a relation of the form

$$A^j x_0 = \sum_{i=j+1}^{q-1} \alpha_i A^i x_i = A^{j+1}(\sum_{i=j+1}^{q-1} \alpha_i A^{i-j-1} x_0) = A^{j+1}y.$$

It follows from the definition of q that

$$A^{q-1}x_0 = A^{q-j-1}A^j x_0 = A^{q-j-1}A^{j+1}y = A^q y = 0;$$

since this contradicts the choice of x_0, we must have $\alpha_j = 0$ for each j.

It is clear that $\mathfrak{H}$ reduces A; to construct $\mathfrak{K}$ we go by induction on the index q of nilpotence. If $q = 1$ the theorem is trivial; we assume the result for $q-1$. The range $\mathfrak{R}$ of A is a linear manifold which reduces A; on $\mathfrak{R}$, A is nilpotent of index $q-1$. We write $\mathfrak{H}_0 = \mathfrak{H} \cap \mathfrak{R}$ and $y_0 = Ax_0$; then $\mathfrak{H}_0$ is spanned by the linearly independent vectors $y_0, Ay_0, \ldots, A^{q-2}y_0$. The induction hypothesis may be applied: $\mathfrak{R}$ is the direct sum of $\mathfrak{H}_0$ and some other linear manifold $\mathfrak{K}_0$, and $A\mathfrak{K}_0 \subset \mathfrak{K}_0$.

We write $\mathfrak{K}_1$ for the set of all vectors x for which Ax is in $\mathfrak{K}_0$; it is clear that $\mathfrak{K}_1$ is a linear manifold. The temptation is great to set $\mathfrak{K} = \mathfrak{K}_1$ and to attempt to prove that this $\mathfrak{K}$ has the desired properties, but unfortunately this need not be true: $\mathfrak{H}$ and $\mathfrak{K}_1$ need not be disjoint. (It is true, although we shall not make explicit use of this fact, that the intersection of $\mathfrak{H}$ and $\mathfrak{K}_1$ is contained in the null space $\mathfrak{N}$ of A). That in spite of this $\mathfrak{K}_1$ is useful is caused by the fact that $\mathfrak{H} + \mathfrak{K}_1 = \mathfrak{V}$. For if x is any vector then Ax is in $\mathfrak{R}$ and consequently $Ax = y + z$ with y in $\mathfrak{H}_0$ and z in $\mathfrak{K}_0$. The general element of $\mathfrak{H}_0$ is a linear combination of $Ax_0, \ldots, A^{q-1}x_0$; hence we have

$$y = \sum_{i=1}^{q-1} \alpha_i A^i x_0 = A(\sum_{i=0}^{q-2} \alpha_i A^i x_0) = Ay_1,$$

where y_1 is in $\mathfrak{H}$. Consequently $Ax = Ay_1 + z$, $A(x - y_1) = z$, so that $A(x - y_1)$ is in $\mathfrak{K}_0$. This means that $x - y_1$ is in $\mathfrak{K}_1$, so that x is the sum of an element (namely y) of $\mathfrak{H}$, and an element of $\mathfrak{K}_1$.

As far as disjointness is concerned we can say at least that $\mathfrak{H} \cap \mathfrak{K}_0 = \mathfrak{O}$. For if x is in $\mathfrak{H}$, then Ax is in $\mathfrak{H}_0$. Since $\mathfrak{K}_0$ is invariant under A, Ax is in $\mathfrak{K}_0$ along with x; consequently if x is common to $\mathfrak{H}$ and $\mathfrak{K}_0$ then Ax is common to $\mathfrak{H}_0$ and $\mathfrak{K}_0$, so that $Ax = 0$. But for an element x of $\mathfrak{H}$, $Ax = 0$ implies that x is in $\mathfrak{H}_0$. (If $x = \sum_{i=0}^{q-1} \alpha_i A^i x_0$, and $Ax = \sum_{i=1}^{q-1} \alpha_{i-1} A^i x_0 = 0$, then, using the linear independence of the $A^j x_0$, it follows that $\alpha_0 = \ldots = \alpha_{q-2} = 0$, so that $x = \alpha_{q-1} A^{q-1} x_0$). Hence if x belongs to $\mathfrak{H} \cap \mathfrak{K}_0$ then it also belongs to $\mathfrak{H}_0 \cap \mathfrak{K}_0$ and consequently $x = 0$.

The situation now is this: $\mathfrak{H}$ and $\mathfrak{K}_1$ together span $\mathfrak{V}$ and $\mathfrak{K}_1$ contains the two disjoint linear manifolds $\mathfrak{K}_0$ and $\mathfrak{H} \cap \mathfrak{K}_1$. If we let $\mathfrak{K}_0'$ be any complement of $\mathfrak{K}_0 \oplus (\mathfrak{H} \cap \mathfrak{K}_1)$ in $\mathfrak{K}_1$, i.e. if $\mathfrak{K}_0' \oplus \mathfrak{K}_0 \oplus (\mathfrak{H} \cap \mathfrak{K}_1) = \mathfrak{K}_1$ then we may write

$\mathfrak{K} = \mathfrak{K}_0' \oplus \mathfrak{K}_0$; we assert that this $\mathfrak{K}$ has the desired properties. In the first place $\mathfrak{K} \subset \mathfrak{K}_1$ and $\mathfrak{K}$ is disjoint from $\mathfrak{H} \cap \mathfrak{K}_1$; it follows that $\mathfrak{H} \cap \mathfrak{K}_1 = \mathfrak{O}$. In the second place $\mathfrak{H} \oplus \mathfrak{K}$ contains both $\mathfrak{H}$ and $\mathfrak{K}_1$, so that $\mathfrak{H} \oplus \mathfrak{K} = \mathfrak{V}$. Finally $\mathfrak{K}$ is invariant under A, since the fact that $\mathfrak{K} \subset \mathfrak{K}_1$ implies that $A\mathfrak{K} \subset \mathfrak{K}_0 \subset \mathfrak{K}$; q.e.d.

Later we shall need the following remark. If $\tilde{x}_0$ is any other vector for which $A^{q-1}\tilde{x}_0 \neq 0$, and if $\tilde{\mathfrak{H}}$ is the linear manifold spanned by the vectors $\tilde{x}_0, \ldots, A^{q-1}\tilde{x}_0$, and if, finally, $\tilde{\mathfrak{K}}$ is any linear manifold which together with $\tilde{\mathfrak{H}}$ completely reduces A, then the behaviour or A on $\tilde{\mathfrak{H}}$ and $\tilde{\mathfrak{K}}$ is the same as its behaviour on $\mathfrak{H}$ and $\mathfrak{K}$ respectively. (In other words: in spite of the apparent non uniqueness in the statement of Theorem 1, everything is uniquely determined up to isomorphisms.) This follows immediately from our discussion of quotient spaces -- this in fact is the reason we introduced them.

Using Theorem 1 we can find a complete geometric characterization of nilpotent transformations.

THEOREM 2. If A is a nilpotent linear transformation of index q on a finite dimentional vector space $\mathfrak{V}$, then we can find a positive integer r, r vectors $x_1, \ldots, x_r$, and r positive integers $q_1, \ldots, q_r$ such that the vectors

$$x_1, Ax_1, \ldots, A^{q_1 - 1}x_1,$$

$$x_2, Ax_2, \ldots, A^{q_2 - 1}x_2$$

$$\cdots$$

$$x_r, Ax_r, \ldots, A^{q_r - 1}x_r,$$

form a basis for $\mathfrak{V}$, and such that $A^{q_1}x_1 = A^{q_2}x_2 = \cdots = A^{q_r}x_r = 0$. The integers $r, q_1, \ldots, q_r$ are a complete set of invariants of A; if, in other words, B is any other nilpotent linear transformation on a finite dimensional vector space $\mathfrak{W}$ then there is an isomorphism T between $\mathfrak{V}$ and $\mathfrak{W}$ for which $TAT^{-1} = B$ if and only if B has the same $r, q_1, \ldots, q_r$ as A.

PROOF. We write $q_1 = q$, and we choose x_1 to be any vector for which $A^{q_1-1}x_1 \neq 0$. From Theorem 1 we know that the linear manifold spanned by $x_1, Ax_1, \ldots, A^{q_1-1}x_1$ completely reduces A, or, in other words, that we may find a complementary manifold which also reduces A and which, naturally, has definitely lower dimension than $\mathfrak{V}$. On this complementary manifold A is nilpotent of index, say, q_2; we apply to this manifold the same reduction procedure (beginning with a vector x_2 for which $A^{q_2-1}x_2 \neq 0$), and we continue thus by induction until we exhaust the space. This proves the existence we asserted; the uniqueness follows from the uniqueness (up to isomorphisms) of the decomposition given by Theorem 1.

Using the basis $\{A^i x_j\}$ the matrix of A takes on a particularly simple form: every matrix element not on the diagonal just below the main diagonal vanishes, (i.e. $\alpha_{ij} \neq 0$ implies $j = i - 1$), and the elements below the main diagonal begin (at top) with a string of 1's followed by a single 0, then go on with another string of 1's followed by a 0, and continue so on to the end, with the lengths of the strings of 1's monotonely decreasing (or, at any rate, non increasing).

It is a sound geometric intuition that makes most

people conjecture that for linear transformations being non singular and being in some sense zero are exactly opposite notions. Our disappointment in finding that $\mathfrak{R}(A)$ and $\mathfrak{N}(A)$ need not be disjoint is connected with this conjecture. The situation can be straightened out by relaxing the sense in which we interpret "being zero"; for most practical purposes a linear transformation some power of which is zero (i.e. a nilpotent transformation) is as zeroish as we can expect it to be. Although we cannot say that a linear transformation is either non singular or "zero" even in the extended sense of zeroness, we can say how any transformation is made up of these two extreme kinds.

THEOREM 3. Every linear transformation A on a finite dimensional vector space $\mathfrak{V}$ is the direct sum of a nilpotent transformation and a non singular transformation.

PROOF. We consider the null space of the k-th power of A: this is a linear manifold $\mathfrak{N}_k = \mathfrak{N}(A^k)$. Clearly $\mathfrak{N}_1 \subset \mathfrak{N}_2 \subset \cdots$. We assert first: if ever $\mathfrak{N}_k = \mathfrak{N}_{k+1}$ then $\mathfrak{N}_k = \mathfrak{N}_{k+j}$ for all positive integers j. For if $A^{k+j}x = 0$ then $A^{k+1}A^{j-1}x = 0$, whence (using the fact that $\mathfrak{N}_k = \mathfrak{N}_{k+1}$) $A^kA^{j-1}x = 0$, so that $A^{k+j-1}x = 0$. In other words $\mathfrak{N}_{k+j}$ is contained in (and therefore equal to) $\mathfrak{N}_{k+j-1}$; induction on j establishes our assertion.

Since $\mathfrak{V}$ is finite dimensional the manifolds $\mathfrak{N}_k$ cannot continue to increase indefinitely; we let q be the smallest positive integer for which $\mathfrak{N}_q = \mathfrak{N}_{q+1}$. It is clear that $\mathfrak{N}_q$ reduces A (in fact each $\mathfrak{N}_k$ does). We write $\mathfrak{R}_k = \mathfrak{R}(A^k)$ for the range of A^k, (so that again it is clear that $\mathfrak{R}_q$ reduces A); we shall prove that $\mathfrak{N}_q \oplus \mathfrak{R}_q = \mathfrak{V}$, and that on $\mathfrak{N}_q$ A is nilpotent, whereas on $\mathfrak{R}_q$ it is non singular.

If x is a vector common to $\mathfrak{N}_q$ and $\mathfrak{R}_q$ then $A^q x = 0$ and $x = A^q y$ for some y. It follows that $A^{2q} y = 0$, and hence, using the definition of q, that $x = A^q y = 0$. Thus we have shown that the range and the null space of A^q are disjoint; a dimensionality argument (see §37) shows that they span $\mathfrak{V}$, so that $\mathfrak{V}$ is their direct sum. It follows from the definitions of q and $\mathfrak{N}_q$ that A on $\mathfrak{N}_q$ is nilpotent of index q. If, finally, x is in $\mathfrak{R}_q$ (so that $x = A^q y$ for some y) and $Ax = 0$ then $A^{q+1} y = 0$, whence $x = A^q y = 0$; this shows that A is non singular on $\mathfrak{R}_q$ and concludes the proof of Theorem 3.

Once again we remind the reader that the consideration of quotient spaces shows that (up to isomorphisms) the behaviour of the nilpotent and non singular parts of A is uniquely determined by A.

We can now use our results on nilpotent transformations to study the structure of arbitrary transformations. The method of getting a nilpotent transformation out of an arbitrary one may seem like a conjuring trick, but it is a very useful trick which is very often employed.

THEOREM 4. If A is any linear transformation on an n-dimensional vector space $\mathfrak{V}$, let $\lambda_1, \ldots, \lambda_p$ be the distinct proper values of A, with respective multiplicities $m_1, \ldots, m_p$. Then $\mathfrak{V}$ is the direct sum of p linear manifolds $\mathfrak{M}_1, \ldots, \mathfrak{M}_p$, of dimensions $m_1, \ldots, m_p$ such that each $\mathfrak{M}_j$ reduces A and such that on $\mathfrak{M}_j$, A has the form $B_j + \lambda_j 1$ where B_j is nilpotent.

PROOF. Take any fixed $j = 1, \ldots, p$, and consider the linear transformation $A_j = A - \lambda_j 1$. To A_j we may apply the decomposition of Theorem 3 to obtain linear manifolds $\mathfrak{M}_j$ and $\mathfrak{N}_j$ such that on $\mathfrak{M}_j$ A_j is nilpotent and on $\mathfrak{N}_j$ it is non singular. Since $\mathfrak{M}_j$ is invariant under A_j it is also invariant under $A_j + \lambda_j 1 = A$. Hence the determinant $\Delta(A - \lambda 1)$ is (for every λ) the product of the two corresponding determinants for the two linear transformations that A becomes when we consider it on $\mathfrak{M}_j$ and $\mathfrak{N}_j$ separately. Since on $\mathfrak{M}_j$ the only proper value of A is λ_j and since on $\mathfrak{N}_j$ A does not have the proper value λ_j ($A - \lambda_j 1$ is non singular on $\mathfrak{N}_j$) it follows that the dimension of $\mathfrak{M}_j$ is exactly m_j, and that for $i \neq j$ $\mathfrak{M}_i$ and $\mathfrak{M}_j$ are disjoint. A dimension argument proves that $\mathfrak{M}_1 \oplus \cdots \oplus \mathfrak{M}_p = \mathfrak{B}$ and thereby concludes the proof of the theorem.

We shall leave to the reader the details of putting together the results of Theorems 2 and 4; we shall merely describe the final result in matricial language.

Given any linear transformation A on a finite dimensional vector space $\mathfrak{B}$ there exists a basis $\mathfrak{X}$ of $\mathfrak{B}$ such that with respect to this basis the matrix of A has the following form. Every element not on or immediately below the main diagonal vanishes. On the main diagonal there appear the distinct proper values of A, each a number of times equal to its multiplicity. Below any particular proper value λ there appear only 1's and 0's, and these in the following way: there are chains of 1's followed by a single 0, with the lengths of the chains decreasing as we read from top to bottom. This matrix is the <u>Jordan</u> or <u>classical canonical form</u> of A; we have $B = TAT^{-1}$ if and only if the classical canonical forms of A and of B are the same except for the order of the proper values. (Thus, in particular, a linear transformation A has in some coordinate system a diagonal matrix if and only if its classical canonical form is already diagonal, i.e. if every chain of 1's has length zero!)

Let us introduce some notation. Let A have p distinct proper values $\lambda_1, \ldots, \lambda_p$ with multiplicities $m_1, \ldots, m_p$ as before; let the number of chains of 1's under λ_j be r_j, and let the lengths of these chains be $q_{j1} - 1, q_{j2} - 1, \ldots, q_{j,r_j} - 1$. The polynomial $(\lambda - \lambda_j)^{q_{j,i}}$ is called an <u>elementary divisor</u> of A of <u>multiplicity</u> $q_{j,i}$ belonging to the proper value λ_j. An elementary divisor is called <u>simple</u> if $q_{j,i} = 1$ (so that the chain length $q_{j,i} - 1 = 0$); we see that a linear transformation A has (in a suitable coordinate system) a diagonal matrix if and only if the elementary divisors are simple.

Theorem 4 does for arbitrary linear transformations what the spectral theorem did for normal ones. We make one application to exhibit the power of the theorem. The linear transformation B_j, described in the statement, is such that $B_j - \lambda_j 1$ is nilpotent of index q_{j1}, or, in other words, B_j is annulled by the polynomial $(\lambda_j - \lambda)^{q_{j1}}$. It follows that A is annulled by the product of these polynomials (i.e. by the product of the elementary divisors of highest multiplicities); this product is called the <u>minimal polynomial</u> of A. It is quite easy to see (since the index of nilpotence of $B_j - \lambda_j 1$ is exactly q_{j1})) that this polynomial is indeed uniquely determined (up to a multiplicative factor) as the polynomial of smallest degree which annuls A. Since the characteristic polynomial $\Delta(A - \lambda 1)$ is the product of all the elementary divisors and therefore a multiple of the minimal polynomial we obtain the <u>Hamilton - Cayley equation</u>: every linear transformation is annulled by its characteristic polynomial.

APPENDIX II

DIRECT PRODUCTS

In this appendix we shall describe a method of putting two vector spaces together to make a third, namely the formation of their direct product, $\mathfrak{W} = \mathfrak{U} \oplus \mathfrak{V}$ Although we had no occasion to make use of direct products in this book their theory is closely allied to some of the subjects we did treat and is useful in other related parts of mathematics, such as the theory of group representations and the tensor calculus. The notion is essentially more complicated than that of direct sum; we shall therefore begin by giving some examples of what a direct product should be, and the study of these examples will guide us in laying down the definition.

(1) Let $\mathfrak{U}$ be the set of all polynomials $x(s)$ in a real variable s; let $\mathfrak{V}$ be the set of all polynomials in another real variable t; and, finally, let $\mathfrak{W}$ be the set of all polynomials $z(s,t)$ in the two real variables s and t. $\mathfrak{U}$, $\mathfrak{V}$, and $\mathfrak{W}$ are all vector spaces; in this particular case we should like to call $\mathfrak{W}$, or something like it, the direct product of $\mathfrak{U}$ and $\mathfrak{V}$. One reason for this terminology is that if we take any x in $\mathfrak{U}$ and any y in $\mathfrak{V}$, we may form their product $x(s)y(t) = z(s,t)$. (This is the ordinary numerical product of two polynomials). Clearly $z(s,t)$ is an element of $\mathfrak{W}$. (Here, as before, we are studiously ignoring the irrelevant fact that we may even multiply together elements of $\mathfrak{U}$, i.e. that the product of two polynomials is another one. Vector spaces in which a decent concept of multiplication is defined are called

algebras and their study, as such, lies outside the scope of this book).

(2) In the preceding example we considered vector spaces whose elements are functions of real variables. We may if we desire view the simple vector space $\mathfrak{C}_n$ as the collection of all functions defined on a set consisting of exactly n points, say the first n positive integers. In other words a vector $\{\xi_1, \ldots, \xi_n\}$ may be considered as a function $\xi(i)$ of i, defined for $i = 1, \ldots, n$; the definition of the vector operations in $\mathfrak{C}_n$ is such that they correspond, in the new notation, to the ordinary numerical operations performed on the functions $\xi(i)$. If, simultaneously, we consider $\mathfrak{C}_m$ as the collection of all functions, say $\eta(j)$, defined for $j = 1, \ldots, m$, then we should like the direct product of $\mathfrak{C}_n$ and $\mathfrak{C}_m$ be the set of all functions $\zeta(i,j)$ defined for $i=1, \ldots, n$, $j = 1, \ldots, m$. The direct product, in other words, is the collection of all functions defined on a set consisting of exactly mn points: this is $\mathfrak{C}_{mn}$. This example brings out a property of direct products, namely the multiplicativity of dimension, that we should like to retain in the general case.

Let us now try to abstract the most important properties of these examples. The definition of direct sum was one possible rigorization of the crude intuitive idea of writing down, formally, the sum of two vectors belonging to different vector spaces. Similarly our examples suggest that the direct product, $\mathfrak{W} = \mathfrak{U} \otimes \mathfrak{V}$, of two vector spaces should be such that to every x in $\mathfrak{U}$ and y in $\mathfrak{V}$ there corresponds a "product", say z, in $\mathfrak{W}$, $z = x \otimes y$, in such a way that the correspondence between x and z, for a fixed y, as well as the correspondence between y and z, for a fixed x, is linear. (I.e. $(\alpha_1 x_1 + \alpha_1 x_2) \otimes y$ should be equal to $\alpha_1(x_1 \otimes y) + \alpha_2(x_2 \otimes y)$, and a similar formula should hold for $x \otimes (\alpha_1 y_1 + \alpha_2 y_2)$). In one word: $x \otimes y$ should be a <u>bilinear</u> (vector valued) function of x and y.

The notion of formal multiplication also suggests that if x' and y' are linear functionals on $\mathfrak{U}$ and $\mathfrak{V}$ then it is their product, $x'(x)\cdot y'(y)$, that should in some sense be the general element of $\mathfrak{W}'$. This product is a function $z'(x,y)$ defined for x in $\mathfrak{U}$ and y in $\mathfrak{V}$, with the property that for each fixed value of one variable it is a linear function of the other: in one word $z'(x,y)$ is a bilinear (scalar valued) functional of x and y.

After one more word of explanation we shall be ready to give the definition. It turns out to be technically preferable to define not $\mathfrak{W}$ itself, but $\mathfrak{W}'$; we shall use the reflexivity property ($\mathfrak{W}'' = \mathfrak{W}$) to define $\mathfrak{W}$. Since we have proved the validity of this property for finite dimensional spaces only, we shall frame the definition for such spaces; we merely remark that the definition may be used in any case (not necessarily finite dimensional) in which reflexivity has been proved.

The preceding discussion was given to motivate the formal work we now begin. The actual definition is not as terrifying as the introduction might seem to make it.

> DEFINITION. If $\mathfrak{U}$ and $\mathfrak{V}$ are any two finite dimensional vector spaces, we write $\mathfrak{W}'$ for the vector space of all bilinear functionals $z'(x,y)$, defined for x in $\mathfrak{U}$ and y in $\mathfrak{V}$. The conjugate space $\mathfrak{W}$ of $\mathfrak{W}'$ is the <u>direct product</u> of $\mathfrak{U}$ and $\mathfrak{V}$, $\mathfrak{W} = \mathfrak{U} \otimes \mathfrak{V}$. To each pair of vectors, x,y, with x in $\mathfrak{U}$ and y in $\mathfrak{V}$, we make correspond a vector z in $\mathfrak{W}$, called the <u>product</u> of x and y, $z = x \otimes y$, defined by $z(z') = z'(x,y)$ for every z' in $\mathfrak{W}'$.

The notation of this definition is slightly out of tune with our preceding custom. $\mathfrak{W}'$ is defined directly, and not as the conjugate space of anything, and doesn't therefore, merit the prime. $\mathfrak{W}$ on the other hand is the conjugate space of something, namely $\mathfrak{W}'$, and should therefore be denoted by $(\mathfrak{W}')'$. The situation is saved by the reflexivity of $\mathfrak{W}'$; since we have the natural isomorphism between $\mathfrak{W}'$ and $(\mathfrak{W}')''$, $\mathfrak{W}'$ may indeed be thought of as the conjugate space of $\mathfrak{W} = \mathfrak{W}''$, and since it is the space $\mathfrak{W}$, and not its conjugate, that we are primarily interested in, we reserve the simpler notation for it.

To get at the point as quickly as possible we slid over a little too lightly a part of the definition. It is probably clear that the set of all bilinear functionals does form a vector space, with the linear operations defined in a way which should by now be obvious to the reader; it takes however another minute's reflection to see that it is finite dimensional. (We referred in the preceding paragraph to the fact that it's reflexive.) A presentation of the detailed theory of bilinear functionals would take us too far afield, and would at the same time be a boring repetition of what we have already done for simple linear functionals. We shall sketch, in the proof of Theorem 1 below, a few of the main facts.

We are not going to go deeply into the theory of direct products. The definition we gave is one of the quickest rigorous approaches to the theory, although it leads to some unpleasant technical difficulties later. Whatever its disadvantages, however, we observe that it obviously has one of the desired properties: it is clear, namely, that $z = x \otimes y$ depends linearly on each of its factors.

Another possible (and quite popular) definition of direct product is by formal products: $\mathfrak{W}$ is the set of all symbols of the form $\sum_i \alpha_i(x_i \otimes y_i)$. (For the purist: $x_i \otimes y_i$ is supposed, in this definition, to stand for the pair $\{x_i, y_i\}$; the multiplication sign is merely a reminder of what to expect.) Neither definition is simple; we adopted the one we gave because it seemed more constructive.

We prove now the two fundamental theorems which had better be true, and which serve, in part, as further justification of the product terminology.

THEOREM 1. The dimension of the direct product, $\mathfrak{W} = \mathfrak{U} \otimes \mathfrak{V}$, of two finite dimensional vector spaces is the product of their dimensions.

PROOF. Since $\mathfrak{W}$ is defined as the conjugate space of an auxiliary space $\mathfrak{W}'$, it is sufficient to prove that if $\mathfrak{U}$ is n-dimensional and $\mathfrak{V}$ is m-dimensional then the dimension of $\mathfrak{W}'$ is nm. We sketch the steps.

(1) If $\mathfrak{X} = (x_1, \ldots, x_n)$ and $\mathfrak{Y} = (y_1, \ldots, y_m)$ are any two bases in $\mathfrak{U}$ and $\mathfrak{V}$ respectively, and if α_{ij}, $i = 1, \ldots, n$, $j = 1, \ldots, m$, is any set of n^2 scalars, then there is one and only one bilinear functional $z'(x,y)$(i.e. there is one and only one element z' in $\mathfrak{W}'$) for which $z'(x_i, y_j) = \alpha_{ij}$. (See Theorem 1, §14). For if $x = \sum_i \xi_i x_i$ and $y = \sum_j \eta_j y_j$ then

(1) $$z'(x,y) = \sum_i \sum_j \xi_i \eta_j z'(x_i, y_j),$$

so that z' is clearly uniquely determined by the prescribed conditions; that any z' is determined at all follows by writing $z'(x_i, y_j) = \alpha_{ij}$ and reading formula (1) from right to left, i.e. defining z' by it.

(2) Using the result just obtained we define $z'_{pq}(x,y)$, for $p = 1, \ldots, n, q = 1, \ldots, m$, by $z'_{pq}(x_i, y_j) = \delta_{ip}\ \delta_{jq}$, (see Theorem 2, §14); the z'_{pq} are a basis in $\mathfrak{W}'$. They are linearly independent, since

$$\sum_p \sum_q \alpha_{pq} z'_{pq} = 0$$

implies

$$0 = \sum_p \sum_q \alpha_{pq}\, \delta_{ip}\, \delta_{jq} = \alpha_{ij};$$

and if $z'(x,y)$ is any element of $\mathfrak{W}'$, with $z'(x_i,y_j) = \alpha_{ij}$, then $z'_{pq}(x,y) = \xi_p\, \eta_q$, so that (substituting into (1)).

(2) $$z'(x,y) = \sum_p \sum_q z'_{pq}(x,y)\, \alpha_{pq}.$$

It follows that the z'_{pq} are a basis in $\mathfrak{W}'$; this completes the proof of the theorem.

THEOREM 2. If $\mathfrak{X} = (x_1, \ldots, x_n)$ and $\mathfrak{Y} = (y_1, \ldots, y_m)$ are bases in $\mathfrak{U}$ and $\mathfrak{V}$ respectively, then the set $\mathfrak{Z}$ of vectors $z_{ij} = x_i \otimes y_j$, $i = 1, \ldots, n$, $j = 1, \ldots, m$, is a basis in $\mathfrak{W} = \mathfrak{U} \otimes \mathfrak{V}$

PROOF. In the proof of the preceding theorem we defined a basis (namely the z'_{ij}) in $\mathfrak{W}'$; let $\tilde{z}_{ij}$, in $\mathfrak{W} = (\mathfrak{W}')'$, be the dual basis. We assert that $\tilde{z}_{ij} = z_{ij}$. To prove this assertion we recall that the $\tilde{z}_{ij}$ are linear functionals on $\mathfrak{W}'$ satisfying the conditions

$$[\tilde{z}_{ij}, z'_{pq}] = \delta_{ip}\, \delta_{jq}.$$

Hence

$$\begin{aligned} \tilde{z}_{ij}(z') = [\tilde{z}_{ij}, z'] &= [\tilde{z}_{ij}, \textstyle\sum_p \sum_q \alpha_{pq} z'_{pq}] \\ &= \textstyle\sum_p \sum_q \alpha_{pq} [\tilde{z}_{ij}, z'_{pq}] \\ &= \alpha_{ij} = z'(x_i, y_j). \end{aligned}$$

Remembering now the definition of the symbol $x \otimes y$, we see that $z_{ij} = x_i \otimes y_j$.

For the purpose of giving examples later we consider the spaces $\mathfrak{P}_n$ and $\mathfrak{P}_m$ (see §2,(4)). We leave it as an exercise for the reader to prove that their product $\mathfrak{S} = \mathfrak{P}_n \otimes \mathfrak{P}_m$ is isomorphic to the space of all polynomials $z(s,t)$ in two real variables, with the property that for each fixed x, $z(s,t)$ is of degree

$\leq m$ in t, and for each fixed t, $z(s,t)$ is of degree $\leq n$ in s, in such a way that the direct product $x \otimes y$ of $x = x(s)$ and $y = y(t)$ corresponds to the ordinary product $z = z(s,t) = x(s)y(t)$.

Let us now try to tie up direct products with linear transformations. Let $\mathfrak{U}$ and $\mathfrak{V}$ be n and m dimensional vector spaces, and let A and B be any two linear transformations on $\mathfrak{U}$ and $\mathfrak{V}$ respectively. Let $\mathfrak{W} = \mathfrak{U} \otimes \mathfrak{V}$ be the direct product of $\mathfrak{U}$ and $\mathfrak{V}$; we define a linear transformation C on $\mathfrak{W}$, called the direct product of A and B, $C = A \otimes B$, as follows. We first define a linear transformation C' on $\mathfrak{W}'$ by the relation

$$C'z'(x,y) = z'(Ax,By);$$

we may then define C as the adjoint of C', $C = (C')'$, or, in symbols.

$$C(z[z'(x,y)]) = z[z'(Ax,By)].$$

If, in particular, we apply C to an element z_0 of the form $z_0 = x_0 \otimes y_0$ (i.e. $z_0(z') = z'(x_0,y_0)$ for every z' in $\mathfrak{W}'$) we obtain $Cz_0 = z_0[z'(Ax_0,By_0)]$, i.e.

(3) $$Cz_0 = Ax_0 \otimes By_0.$$

Since there are quite a few elements in $\mathfrak{W}$ of the form $x \otimes y$, enough at any rate to form a basis, (Theorem 2,), this last relation characterizes C.

The formal rules for operating with direct products are the following:

(4) $$0 \otimes A = A \otimes 0 = 0,$$

(5) $$1 \otimes 1 = 1,$$

(6) $$(A_1 + A_2) \otimes B = (A_1 \otimes B) + (A_2 \otimes B),$$

(7) $$A \otimes (B_1 + B_2) = (A \otimes B_1) + (A \otimes B_2),$$

(8) $$\alpha A \otimes \beta B = \alpha\beta(A \otimes B),$$

(9) $$(A \otimes B)^{-1} = A^{-1} \otimes B^{-1},$$

(10) $$(A_1A_2) \otimes (B_1B_2) = (A_1 \otimes B_1)(A_2 \otimes B_2).$$

Formula (9), like all formulae involving inverses, has to be read with caution: we assert that the existence of A^{-1} and B^{-1} implies the existence of $(A \otimes B)^{-1}$, and the validity of (7); conversely, in a finite dimensional space, the existence of $(A \otimes B)^{-1}$ implies that of A^{-1} and B^{-1}. We shall prove (9) and (10), in reverse order.

Formula (10) follows from the characterization (3) of direct products and the following computation:

$$(A_1A_2 \otimes B_1B_2)(x \otimes y) = (A_1A_2x) \otimes (B_1B_2y)$$
$$= (A_1 \otimes B_1)(A_2x \otimes B_2y) = (A_1 \otimes B_1)(A_2 \otimes B_2)(x \otimes y).$$

As an immediate consequence of (10) we obtain

(11) $$A \otimes B = (A \otimes 1)(1 \otimes B) = (1 \otimes B)(A \otimes 1).$$

To prove (9), suppose that A^{-1} and B^{-1} exist, and form $A \otimes B$ and $A^{-1} \otimes B^{-1}$. Since, by (10), the product of these two transformations, in either order, is 1, $A \otimes B$ has an inverse and this inverse is equal $A^{-1} \otimes B^{-1}$. Conversely suppose that $(A \otimes B)^{-1}$ exists; remembering that we defined direct product for finite dimensional spaces only, we may invoke Theorem 2 of §24: we shall show that $Ax = 0$ implies that $x = 0$, and $By = 0$ implies that $y = 0$. We use (1): $Ax \otimes By = (A \otimes B)(x \otimes y)$. If either factor on the left is zero then $(A \otimes B)(x \otimes y) = 0$, whence $x \otimes y = 0$, so that either $x = 0$ or $y = 0$. Since (by (4)) $B=0$ is impossible, we may find a vector y for which $By \neq 0$. Applying the above argument to this y, together with any x for which $Ax = 0$, it follows that $x = 0$. The same argument with the roles of A and B interchanged proves that B has an inverse.

For an interesting example of the direct product of two transformations we take $\mathfrak{U}$ and $\mathfrak{V}$ to be $\mathfrak{P}_n$ and $\mathfrak{P}_m$, and A and B to be differentiation on $\mathfrak{P}_n$ and $\mathfrak{P}_m$ respectively. Since the space $\mathfrak{U} \otimes \mathfrak{V}$ may be thought of as a space of polynomials in two variables, the direct product $C = A \otimes B$, applied to $z = z(s,t)$, yields the mixed partial derivative $Cz = \frac{\partial^2 z}{\partial s \partial t}$. Proof?

An interesting (and complicated) side of the theory of direct products is the theory of Kronecker products of matrices. Let $\mathfrak{X} = (x_1, \ldots, x_n)$ and $\mathfrak{Y} = (y_1, \ldots, y_m)$ be bases in $\mathfrak{U}$ and $\mathfrak{V}$, and let $[A] = [A; \mathfrak{X}] = (\alpha_{ij})$ and $[A] = [B; \mathfrak{Y}] = (\beta_{pq})$ be the matrices of A and B. What is the matrix of $A \otimes B$ in the coordinate system $\{x_i \otimes y_p\}$?

To answer this question we must recall the discussion in §25 concerning the arrangement of a basis in some linear order. Since, unfortunately, it is impossible to write down a matrix without committing one's self to an order of rows and columns, we shall be frank about it, and arrange the n times m vectors $x_i \otimes y_p$ in the so called lexicographical order, as follows:

$x_1 \otimes y_1, x_1 \otimes y_2, \ldots, x_1 \otimes y_m, x_2 \otimes y_1, \ldots,$

$$x_2 \otimes y_m, \ldots, x_n \otimes y_1, \ldots, x_n \otimes y_m.$$

We may also carry out the following computation:

$$(12) \qquad (A \otimes B)(x_j \otimes y_q) = Ax_j \otimes By_q$$

$$= (\textstyle\sum_i \alpha_{ij} x_i \otimes (\sum_p \beta_{pq} y_p)$$

$$= \textstyle\sum_i \sum_p \alpha_{ij}\, \beta_{pq} (x_i \otimes y_p).$$

This process indicates exactly how far we can get without ordering the basis elements; if, for example, we agree to index the elements of a matrix not with a pair of integers but with a pair of pairs, (i,p) and (j,q), then we know now that the element in the (i,p)-th row and (j,q)-th column is $\alpha_{ij}\, \beta_{pq}$. If we use the lexicographical ordering, the matrix of $A \otimes B$ has the form:

$$
(13)\quad
\left[\begin{array}{ccc|ccc|c}
\alpha_{11}\beta_{11} & \cdots & \alpha_{11}\beta_{1m} & \alpha_{12}\beta_{11} & \cdots & \alpha_{12}\beta_{1m} & \\
\cdot & \cdot & \cdot & \cdot & \cdot & \cdot & \cdots \\
\alpha_{11}\beta_{m1} & \cdots & \alpha_{11}\beta_{mm} & \alpha_{12}\beta_{m1} & \cdots & \alpha_{12}\beta_{mm} & \\
\hline
 & \cdots & & & \cdots & & \cdots \\
\hline
\alpha_{n1}\beta_{11} & \cdots & \alpha_{n1}\beta_{1m} & & & & \alpha_{nn}\beta_{11} \cdots \alpha_{nn}\beta_{1m} \\
\cdot & \cdot & \cdot & & \cdots & & \cdots \\
\alpha_{n1}\beta_{m1} & \cdots & \alpha_{n1}\beta_{mm} & & & & \alpha_{nn}\beta_{m1} \cdots \alpha_{nn}\beta_{mm}
\end{array}\right]
$$

In a condensed notation whose meaning is clear we may write this matrix as

$$
(14)\quad
\begin{bmatrix}
\alpha_{11}[B] & \alpha_{12}[B] & \cdots & \alpha_{1n}[B] \\
\alpha_{21}[B] & \alpha_{22}[B] & \cdots & \alpha_{2n}[B] \\
\cdots & \cdots & \cdots & \cdots \\
\alpha_{n1}[B] & \alpha_{n2}[B] & \cdots & \alpha_{nn}[B]
\end{bmatrix}
$$

This matrix is known as the Kronecker product of [A] and [B], in this order. The rule for forming it is easy to describe in words: replace each element α_{ij} of the n times n matrix [A] by the m times m matrix $\alpha_{ij}[B]$. If in this rule we interchange the roles of A and B (and consequently interchange n amd m) we obtain the definition of the Kronecker product of [B] and [A]. Query: is there an arrangement of the basis vectors $x_i \otimes y_p$, such that the matrix of $A \otimes B$, referred to the arranged coordinate system, is the Kronecker product of [B] and [A]?

Let us now attempt to introduce a sensible inner product into the direct product of two unitary spaces. It is technically easier to define inner product not in $\mathfrak{W} = \mathfrak{U} \otimes \mathfrak{V}$ but in the auxiliary space $\mathfrak{W}$ and then to apply the general theory of the conjugate space of a unitary space to find an inner product in $\mathfrak{W}$.

If $z' = z'(x,y)$ is any element of $\mathfrak{W}'$, z' may be written as a sum of products of the form $x'(x)y'(y)$, where x' and y' are linear functionals on $\mathfrak{U}$ and $\mathfrak{V}$ respectively. Since $\mathfrak{U}$ and $\mathfrak{V}$ are unitary spaces this implies that $z'(x,y)$ may be written as a finite sum of expressions of the form $(x,x_o)(y,y_o)$; say

(15) $$z'(x,y) = \sum_i (x,x_i)(y,y_i).$$

Hence if z_1' and z_2' are any two elements of $\mathfrak{W}'$ we may write

(16) $$z_1'(x,y) = \sum_i (x,x_{i1})(y,y_{i1}),$$

(17) $$z_2'(x,y) = \sum_j (x,x_{j2})(y,y_{j2}),$$

and we may define

$$(z_1',z_2') = \sum_i \sum_j (x_{j2},x_{i1})(y_{j2},y_{i1}).$$

(The conjugate nature of the relation between vectors and linear functionals on a unitary space again necessitates putting x_{j2} before x_{i1}). Before we can even start to prove that this definition fulfills the conditions of the definition of an inner product, we must prove that it defines (z_1',z_2') independently of the representations of z_1' and z_2' as sums in (16) and (17). To do this we observe that

$$\sum_j (x_{j2},x_{i1})(y_{j2},y_{i1}) = \overline{z_2'(x_{i1},y_{i1})},$$

so that $(z_1',z_2') = \sum_i \overline{z_2'(x_{i1},y_{i1})}$ is independent of the particular representation of z_2'. Since, moreover, in any given representations of z_1' and z_2' it is true that $(z_1',z_2') = \overline{(z_2',z_1')}$, it follows that (z_1',z_2') is also independent of the representation of z_1'.

It is easy to verify that the expression (z_1',z_2') is linear in z_1', conjugate linear in z_2', and Hermitian symmetric. The non trivial part of this discussion is the proof that it is also positive definite, i.e. that $(z',z') > 0$ for all $z' \neq 0$.

Using the representation (1) we have

$$(z',z') = \sum_i \sum_j (x_j,x_i)(y_j,y_i).$$

For any complex numbers $\{\xi_i\}$ we have

$$\sum_i \sum_j (x_j,x_i)\,\bar{\xi}_i\,\xi_j = (\sum_j \xi_j x_j,\ \sum_i \xi_i x_i) = \|\sum_j \xi_j x_j\|^2 \geq 0,$$

so that the matrix whose general element is (x_j,x_i) is non negative. Similarly of course the matrix whose general element is (y_j,y_i) is non negative; it follows from Theorem 2 of §69 that the same is true of the Hadamard product $(x_j,x_i)(y_j,y_i))$. Hence

$$\sum_i \sum_j (x_j,x_i)(y_j,y_i)\,\bar{\xi}_i\,\xi_j \geq 0$$

for every set of complex ξ's; choosing $\xi_i = 1$ for all i, we see that $(z',z') \geq 0$ for all z'.

In order to prove that $(z',z') = 0$ implies $z' = 0$ we proceed as follows. For the expression (z_1',z_2'), which by now has all other properties of an inner product, we may prove Schwarz's inequality just as in §44:

$$|(z_1',z_2')|^2 \leq (z_1',z_1')(z_2',z_2').$$

It follows that the vanishing of (z_1',z_1') implies the vanishing of (z_1',z_2') for all z_2'. Let x_0 and y_0 be arbitrary vectors and take in particular

$$z_2' = z_2'(x,y) = (x,x_0)(y,y_0).$$

The vanishing of (z_1',z_2') implies that

$$0 = (z_1',z_2') = \sum_i (x_0,x_{i1})(y_0,y_{i1}) = z_1'(x_0,y_0);$$

hence $z_1'(x_0,y_0) = 0$ for all x_0 and y_0, so that $z_1' = 0$

This concludes the introduction of an inner product in $\mathfrak{W}'$ -- we denote the unitary space so obtained by $\mathfrak{W}^*$. Applying the results of §50 we obtain an inner product in the conjugate space $\mathfrak{W}$ of $\mathfrak{W}^*$.

It is now easy to prove that the inner product defined in $\mathfrak{W}$ has the property that

$$(x_1 \otimes y_1,\ x_2 \otimes y_2) = (x_1,x_2)(y_1,y_2).$$

We write $z_1 = x_1 \otimes y_1$, $z_2 = x_2 \otimes y_2$. We consider also the elements z_1^* and z_2^* of $\mathfrak{W}^*$ defined by

$$z_1^*(x,y) = (x,x_1)(y,y_1),$$
$$z_2^*(x,y) = (x,x_2)(y,y_2),$$

and we define the elements $\tilde{z}_1$ and $\tilde{z}_2$ of $\mathfrak{W}$ by

$$\tilde{z}_1 = \tilde{z}_1(z^*) = (z^*,z_1^*),$$
$$\tilde{z}_2 = \tilde{z}_2(z^*) = (z^*,z_2^*).$$

For an arbitrary $z_0^* = z_0^*(x,y)$ in $\mathfrak{W}^*$, with the representation

$$z_0^*(x,y) = \sum_i (x,x_i^0)(y,y_i^0)$$

we have

$$\tilde{z}_1(z_0^*) = \sum_i (x_1,x_i^0)(y_1,y_i^0),$$
$$\tilde{z}_2(z_0^*) = \sum_i (x_2,x_i^0)(y_2,y_i^0),$$

whence

$$\tilde{z}_1(z_0^*) = z_0^*(x_1,y_1) = z_1(z_0^*),$$
$$\tilde{z}_2(z_0^*) = z_0^*(x_2,y_2) = z_2(z_0^*).$$

(This is very similar to the proof in §52 of the equality of the two natural correspondences between a unitary space and its conjugate). Hence we have, finally,

$$(z_1,z_2) = (\tilde{z}_1,\tilde{z}_2) = (z_2^*,z_1^*) = (x_1,x_2)(y_1,y_2),$$

as was to be proved.

The last proved fact justifies once more the direct product terminology and describes completely the structure of $\mathfrak{W}$ and its relation to $\mathfrak{U}$ and $\mathfrak{V}$. It follows also that if $\{x_i\}$ and $\{y_p\}$ are orthogonal bases in $\mathfrak{U}$ and $\mathfrak{V}$ respectively then

$$(x_i \otimes y_p,\ x_j \otimes y_q) = (x_i,x_j)(y_p,y_q) = \delta_{ij}\,\delta_{pq},$$

so that the $\{x_i \otimes y_p\}$ form an orthonormal set in $\mathfrak{W}$. Since we have already seen that they form a linear basis, it follows that they are a complete orthonormal set, or an orthogonal basis, in $\mathfrak{W}$.

APPENDIX III

HILBERT SPACE

Probably the most useful and certainly the best developed generalization of the theory of finite dimensional unitary spaces to infinite dimensions is the theory of Hilbert space. Without going into details and entirely without proofs we shall now attempt to indicate how this generalization proceeds and what are the main difficulties which have to be overcome.

The definition of Hilbert space is easy: it is an infinite dimensional unitary space satisfying one extra condition. That this condition -- namely completeness -- is automatically satisfied in the finite dimensional case is proved in elementary analysis. In the infinite dimensional case it may be possible that for a sequence $\{x_n\}$ of vectors $\| x_n - x_m \| \longrightarrow 0$ as $n,m \longrightarrow \infty$, but still there is no vector x for which $\| x - x_n \| \longrightarrow 0$; the only effective way of ruling out this possibility is explicitly to assume its opposite. In other words: a Hilbert space is a complete infinite dimensional unitary space. (Sometimes the concept of Hilbert space is restricted by an additional cardinal number condition -- namely separability -- but in recent years, ever since the realization that this additional restriction doesn't pay for itself in results, it has become customary to use "Hilbert space" for the concept we defined.)

It is easy to see that the space $\mathfrak{P}$ of polynomials with the inner product $(x,y) = \int_0^1 x(t)\overline{y(t)}dt$ is not complete. In connection with the notion of completeness of certain particular Hilbert spaces there is a quite extensive mathematical lore: the main assertion of the celebrated Riesz - Fischer theorem is that the space $\mathfrak{F}$ of all functions $x(t)$ for which $|x(t)|^2$ is Lebesgue integrable in the interval $(0,1)$ is (with the same formal definition of inner product as for polynomials) a Hilbert space. Another popular Hilbert space, reminiscent in its appearance of finite dimensional Euclidean space, is the space $\mathfrak{S}$ of all sequences $\{\xi_n\}$ of complex numbers for which $\sum_n |\xi_n|^2$ converges.

Using completeness in order to discuss intelligently the convergence of some infinite series one can proceed for quite some time in building the theory of Hilbert spaces without meeting any difficulties due to infinite dimensionality. Thus our proof of Schwarz's inequality is valid in the most general case and the notions of orthogonality and complete orthonormal sets can be defined exactly as we defined them. Even our proof of Bessel's inequality and of the equivalence of the various possible formulations of completeness for an orthonormal set have to undergo only slight verbal changes. (The convergence of the various infinite series that enter is an automatic consequence of Bessel's inequality). Finally the proof of the existence of complete orthonormal sets parallels closely the proof in the finite case; in the unconstructive proof transfinite induction (or Zorn's lemma) replaces ordinary induction, and even the constructive steps of the Gram - Schmidt process are easily carried out.

In the discussion of manifolds, functionals, and transformations the situation becomes uncomfortable if we do not make a concession to the topology of Hilbert space. Good analogs of all our statements for the finite dimensional case can be proved if we consider <u>closed</u> linear manifolds, <u>continuous</u> linear functionals, and <u>bounded</u> linear transformations. (In a finite dimensional space every linear manifold is closed, every linear functional is continuous, and every linear transformation is bounded.) If, however, we do agree to make these concessions then once more we can coast on our finite dimensional proofs without any change most of the time and with only the insertion of an occasional ϵ the rest of the time. Thus once more we obtain that $\mathfrak{D} = \mathfrak{M} \oplus \mathfrak{M}^{\perp}$ and $\mathfrak{M} = \mathfrak{M}^{\perp\perp}$ and that every linear functional of x has the form (x,y); our definitions of Hermitian and non negative transformations still make sense, and all our theorems about perpendicular projections (as well as their proofs) carry over without change.

The first hint of how things can go wrong comes from the study of unitary transformations. We still call a transformation U unitary if $UU^* = U^*U = 1$, and it is still true that a unitary transformation is isometric, i.e. $\| Ux \| = \| x \|$ for all x, or equivalently $(Ux,Uy) = (x,y)$ for all x and y. It is, however, easy to construct an isometric transformation which is not unitary; because of its importance in the construction of counter examples we shall describe one such transformation. We consider a Hilbert space in which there is a countable complete orthonormal set, say $x_1, x_2, \ldots$. A unique bounded linear transformation U is defined by the conditions $Ux_n = x_{n+1}$ for $n = 1,2, \ldots$; this U is isometric but not unitary since $U^*U = 1$, but $UU^*x_1 = 0$. The theory of Cayley transforms is affected by the distinction between unitary and isometric, but only to a comparatively slight extent -- we omit the description of the differences.

It is when we come to the spectral theory that the whole flavor of the development changes radically. The definition of proper value as a number λ for which $Ax = \lambda x$ has a solution $x \neq 0$ still makes sense and is still useful in many contexts, and our theorem about the reality of the proper values of a Hermitian operator is still true. The notion of proper value loses, however, much of its significance. Proper values are so very useful in the finite dimensional case because they are a handy way of describing the fact that something goes wrong with the inverse of $A - \lambda 1$, and the only thing that can go wrong is that the inverse does not exist. Essentially different things can happen in the infinite dimensional case; just to illustrate the possibilities we mention that, for example, the inverse of $A - \lambda 1$ may exist but be unbounded. That there is no useful generalization of determinant, and hence of the characteristic equation, is the least of our worries. The whole theory has, in fact, attained its full beauty and maturity only after the slavish imitation of such finite dimensional methods was given up.

After some appreciation of the fact that the infinite dimensional case has to overcome great difficulties, it comes as a pleasant surprise that the spectral theorem for Hermitian (and even for normal) operators does have a very beautiful and powerful analog. (Although we describe the theorem for bounded operators only there is a large class of unbounded ones for which it is valid). In order to be able to understand the analogy let us re-examine the finite dimensional case.

Let A be a Hermitian linear transformation on a finite dimensional vector space and let $A = \sum_j \lambda_j F_j$ be its spectral form. Let us write $E(\lambda) = \sum_{\lambda_j < \lambda} F_j$, where the summation is extended over those values of j for which $\lambda_j < \lambda$. It is clear that, for each λ, $E(\lambda)$ is a (perpendicular) projection and it is easy to see that the characteristic properties of the $E(\lambda)$ are the following.

(1) For $\lambda < \mu$, $E(\lambda) \leq E(\mu)$.

(2) For λ sufficiently large $E(\lambda) = 1$ and $E(-\lambda) = 0$.

(3) $AE(\lambda) = E(\lambda)A$ for all λ.

(4) For $\epsilon > 0$ sufficiently small, $E(\lambda - \epsilon) = E(\lambda)$.

(5) $A = \sum_j \lambda_j [E(\lambda_{j+1}) - E(\lambda_j)]$.

Those familiar with Stieltjes integration will recognize the sum in (5) as a typical approximating sum to an integral of the form $\int \lambda \, dE(\lambda)$ and will therefore see how one may expect the generalization to go. The exact statements (4) and (5) are replaced by the limiting statements,

(4') $\lim_{\epsilon \to 0} E(\lambda - \epsilon) = E(\lambda)$, $\epsilon > 0$;

(5') $A = \int_{-\infty}^{+\infty} \lambda \, dE(\lambda)$.

Except for this obvious alteration the spectral theorem remains true in Hilbert space. We have, of course, to interpret correctly the meaning of the limiting operations. Once more we are faced with the three possibilities we mentioned in §75 (called uniform, strong, and weak convergence respectively); it turns out that (4') is to be given the strong and (5') the uniform interpretation. (The reader deduces of course from our language that the three possibilities are indeed distinct in Hilbert space.)

We have seen that the projections F_j entering into the spectral form of A in the finite dimensional case are very simple functions of A. (§65). Since the $E(\lambda)$ are obtained from the F_j by summation they also are functions of A, and it is quite easy to describe what functions. We write $g_\lambda(\tau) = 1$ if $\tau \leqq \lambda$ and $g_\lambda(\tau) = 0$ otherwise; then $E(\lambda) = g_\lambda(A)$. This fact gives the main clue to the proof of the spectral theorem in Hilbert space. The usual process is to discuss the functional calculus for polynomials and by limiting processes to extend it to functions of Baire class 1, (i.e. to such functions as $g_\lambda(\tau)$.) Once this is done we may write, for any given A, $E(\lambda) = g_\lambda(A)$ by definition; there is no particular difficulty in the proof of assertions (1), (2), (3), (4'), (5').

After the spectral theorem is proved it is easy to

deduce from it the analogs of our theorems concerning square roots, the general functional calculus, the polar decomposition, and properties of commutativity, and in fact to answer practically every askable question concerning bounded normal operators.

The chief difficulties that remain are the considerations of non normal and of unbounded operators. Concerning non normal transformations it is easy to describe the state of our knowledge -- it is non existent. No even unsatisfactory analog exists of the superdiagonal form or of the Jordan canonical form and the theory of elementary divisors. Very different is the situation concerning normal (and particularly Hermitian) unbounded transformations. (The reader will sympathize with the desire to treat such transformations if he recalls that the first and most important functional operation that most of us learn is differentiation.) In this connection we shall barely hint at the main obstacle the theory faces. It is not very difficult to show that if a Hermitian linear transformation A is defined for all vectors of Hilbert space then it is bounded. In other words the first requirement concerning transformations that we are forced to give up is that they be defined everywhere. The discussion of the precise domain on which a Hermitian transformation may be defined and of the extent to which this domain may be enlarged is the chief new difficulty encountered in the study of unbounded operators -- for the details we invite the reader to consult the texts listed in the bibliography.

BIBLIOGRAPHY

ALBERT, A. A., Modern Higher Algebra, Chicago, 1937.

BANACH, S., Théorie des Opérations Linéaires, Warszawa, 1932.

COURANT, R. and HILBERT, D., Methoden der Mathematischen Physik, volume 1, Berlin, 1931.

FRAZER, R. A., DUNCAN, W. J., and COLLAR, A. R., Elementary Matrices, Cambridge, 1938.

HASSE, H., Höhere Algebra, volume 1, Berlin, 1933.

MacDUFFEE, C. C., The Theory of Matrices, Berlin, 1933.

MURNAGHAN, F. D., The Theory of Group Representations, Baltimore, 1938.

von NEUMANN, J., Mathematische Grundlagen der Quantenmechanik, Berlin, 1932.

SCHREIER, O. and SPERNER, E., Vorlesungen über Matrizen, Leipzig, 1932.

STONE, M. H., Linear Transformations in Hilbert Space, New York, 1932.

van der WAERDEN, B. L., Moderne Algebra, two volumes, Berlin, 1937 - 1940.

WEDDERBURN, J. H. M., Lectures on Matrices, New York, 1934.

WEYL, H., The Classical Groups, Princeton, 1939.

WINTNER, A., Spektraltheorie der Unendlichen Matrizen, Leipzig, 1929.

LIST OF NOTATIONS

Throughout this book we have observed, whenever possible, the following conventions. Lower case Greek letters, α, β, γ, λ, μ, ν, σ, τ, stand for scalars in general and real and complex numbers in particular; lower case Latin letters around the middle of the alphabet, i,j,k,m,n,p,q,r, are used for positive integers, and those at the end of the alphabet, x,y,z,u,v, for vectors and linear functionals. Capital Latin letters, A,B,C,E,F,G,S,T,U,V, denote linear transformations and capital German letters, $\mathfrak{H}$, $\mathfrak{K}$, $\mathfrak{M}$, $\mathfrak{N}$, $\mathfrak{P}$, $\mathfrak{Q}$, $\mathfrak{R}$, $\mathfrak{S}$, $\mathfrak{U}$, $\mathfrak{V}$, $\mathfrak{W}$, $\mathfrak{X}$, $\mathfrak{Y}$, stand for vector spaces, linear manifolds, and sets of vectors in general. The most notable violations of these rules are our occasional use of s,t for real variables, f,g,h,p,q,r for polynomials and other functions, T for trace, R and I for real and imaginary part, and i for $\sqrt{-1}$. The capital Greek letters $\sum$ and $\prod$ are reserved as usual for addition and multiplication. In general we indicate summation by a symbol such as $\sum_i \alpha_{ij}x_i$; this is to be interpreted as summation over the entire range of the index i of $\sum$. Only when we depart from this convention do we use such an expression as $\sum_{j=1}^{m} y_j$.

Superscripts, as in A^n, λ^n, generally stand for exponents; on some rare occasions (notably in §69) when the alphabet is nearly exhausted we use them merely as indices.

The braces $\{\ \ldots\ \}$ are generally used to denote a set; thus $\{\xi_1, \ldots, \xi_n\}$ may be the coordinates of a vector, and $\{x_n\}$ may stand for a finite or infinite sequence of vectors. To this rule there is one important exception; when we wish explicitly to write out a finite set of vectors we use the parentheses $(x_1, \ldots, x_n)$ instead of the braces in order to emphasize the fact that we are not discussing coordinates.

We use the double arrow $\Longrightarrow$ for implication and $\rightleftarrows$ for equivalence (i.e. implication in both directions); the simple arrow $\longrightarrow$ denotes the effect of a mapping or, sometimes, the convergence of a sequence -- the correct interpretation will always be clear from the context.

We adjoin a list of some of the other symbols we used; the numbers refer to sections and Roman numerals to the appendices.

A^{-1}	24
A'	32
$[A]$	25
$[A; \mathfrak{X}]$	25
A^*	51
D	20
δ_{ij}	6
Δ	39
$\mathfrak{C}_n$	2
inf	71
$\mathfrak{H} + \mathfrak{K}$	10
J	20
$\mathfrak{M}^0$	16
$\mathfrak{M}^{\perp}$	46
$\mathfrak{N}(A)$	36
$\nu(A)$	37
$\mathfrak{O}$	9
$\mathfrak{P}$	2
$P_{\mathfrak{M}}$	57
$\mathfrak{P}_n$	2
$\mathfrak{R}(A)$	36
$\mathfrak{R}_n$	2

INDEX OF DEFINITIONS

(Numbers refer to sections; Roman numerals to appendices.)

Lightning Source UK Ltd.
Milton Keynes UK
UKHW040619200522
403285UK00001B/70

9 780691 090955